ALLEN COUNTY PUBLIC LIBRARY

FORT WAYNE, INDIANA 46802

You may return this book to any agency, branch,
or bookmobile of the Allen County Public Library.

DEMCO

OTHER BOOKS OF INTEREST
FROM COMPUTER SCIENCE PRESS

Lowell A. Carmony, Robert J. McGlinn, Ann Miller Millman, and Jerry P. Becker
Apple Pascal: A Self-Study Guide for the Apple II Plus, IIe, and IIc

Lowell A. Carmony and Robert L. Holliday
Macintosh Pascal

Narain Gehani
Advanced C: Food of the Educated Palate

Narain Gehani
C for Personal Computers: IBM PC, AT&T PC 6300, and Compatibles

Owen Hanson
Design of Computer Data Files

Marvin K. Simon, Jim K. Omura, Robert A. Scholtz, and Barry K. Levitt
Spread Spectrum Communications, Volumes I, II, III

William W. Wu
Elements of Digital Satellite Communication, Volumes I and II

Personal Computers
&
Data Communications

DIMITRIS N. CHORAFAS

COMPUTER SCIENCE PRESS

Computer Science Press, Inc.
1803 Research Boulevard
Rockville, Maryland 20850

1 2 3 4 5 6   91 90 89 88 87 86

Library of Congress Cataloging in Publication Data
Chorafas, Dimitris N.
  Personal computers and data communications.

  1. Data transmission systems.   2. Microcomputers.
I. Title.
TK5105.C53   1985   384   84-19972
ISBN 0-88175-052-2

CONTENTS

2271204

PREFACE .... **x**

PART ONE: A SYSTEM AT WORK

CHAPTER 1   **PERSONAL WORKSTATIONS** **3**
  1.1 Introduction 3
  1.2 Defining and Setting Up the Workstation 4
  1.3 Software/Hardware Trends and Technologies 8
  1.4 Interfacing to the End User 12
  1.5 Connecting Personal Computers and Mainframes .. 16

CHAPTER 2   **THE OFFICE OF THE FUTURE** **19**
  2.1 Introduction 19
  2.2 Direction in Office Systems Development 20
  2.3 Background for the New Technologies 23
  2.4 New Applications 26

CHAPTER 3   **NEW KNOW-HOW** **31**
  3.1 Introduction 31
  3.2 Trends in Information Technology 32
  3.3 A Transition in Applications Images 35
  3.4 Document Handling and the Database 38
  3.5 Capitalizing on Know-How 41

CHAPTER 4   **PERSPECTIVES IN INFORMATION SYSTEMS** .. **44**
  4.1 Introduction 44
  4.2 A Truly Distributed System 45
  4.3 Cost Effectiveness Made Possible by Technology .. 47
  4.4 What Makes a Good Product? 48
  4.5 Insight and Foresight in Computer Decisions 50
  4.6 Telling the Machine What To Do 53

CHAPTER 5   **PERSONAL COMPUTING FOR SENIOR**
                MANAGEMENT **57**
  5.1 Introduction 57
  5.2 Teaching the Man with the Problem 58

v

5.3 The Talent Market 61
5.4 A Lifelong Learning Experience 65
5.5 Top Management's Workstation 67

PART TWO: PERSONAL COMPUTING

CHAPTER 6   PROMOTING OFFICE AUTOMATION **73**
6.1 Introduction 73
6.2 The System is for the End User 74
6.3 Systems Studies 77
6.4 Emphasis on Communications 79
6.5 Artificial Intelligence Today 83

CHAPTER 7   INFRASTRUCTURE FOR EXPERT SYSTEMS **87**
7.1 Introduction 87
7.2 A System to Enlarge the Mind's Capabilities 88
7.3 Problems and Solutions 90
7.4 Learning from Experience 92
7.5 Bordering on Imagination 94
7.6 Management by Objectives 96

CHAPTER 8   PREPARING FOR DECISION SUPPORT **101**
8.1 Introduction 101
8.2 Looking at DSS in a Productive Way 101
8.3 Choosing Candidates for DSS 104
8.4 Retrieving the Infopages 108
8.5 Keeping the Options Open 111

CHAPTER 9   USING THE SPREADSHEET **114**
9.1 Introduction 114
9.2 Initialization of the Machine 115
9.3 Commands at the User Level 115
9.4 EDIT Command 120
9.5 Format Definition for the Spreadsheet 121
9.6 Using Infopages 122

CHAPTER 10   DISTRIBUTED SYSTEMS IN MANUFACTURING
AND MERCHANDIZING
10.1 Introduction 124
10.2 Functional Description 125
10.3 Keeping Cost-Effectiveness Under Perspective ... 129
10.4 Sense of Responsibility 132
10.5 An Industrial Network Base 134

PART THREE: USING ADVANCED TECHNOLOGY

CHAPTER 11 CHALLENGES WITH SEMICONDUCTORS 141
    11.1 Introduction 141
    11.2 Microprocessors 142
    11.3 Design and Manufacturing 145
    11.4 The Silicon Compiler 147
    11.5 Price/Performance 148

CHAPTER 12 BIOTECHNOLOGY AND THE BIOCHIP 153
    12.1 Introduction 153
    12.2 What is Biotechnology? 154
    12.3 A Steady Progress in Electronics 156
    12.4 The New Microprocessors 160
    12.5 Biotechnology and Health Care 163
    12.6 Impact in the Agribusiness 165

CHAPTER 13 INTERFACING THE MAN TO THE MACHINE .. 168
    13.1 Introduction 168
    13.2 Functional Commands, Cursor, Mouse,
        and Joystick 169
    13.3 Implementing Softcopy Windows 173
    13.4 Touch Sensing Screens 175
    13.5 Optimizing Display Characteristics 178

CHAPTER 14 VOICE INPUT/OUTPUT 180
    14.1 Introduction 180
    14.2 The Voice Handling Effort 181
    14.3 Progress in Voice Recognition 184
    14.4 Focusing on Voice Input 187
    14.5 Voice Response 190

CHAPTER 15 DISC STORAGE 194
    15.1 Introduction 194
    15.2 Winchester Technology 195
    15.3 Removable Discs 199
    15.4 Flexible Discs 200
    15.5 Vertical Recording 202
    15.6 Streaming Tape for Disc Backup 203
    15.7 Optical Discs 206

CHAPTER 16 BENCHMARKING THE PC 208
    16.1 Introduction 208
    16.2 The Range of Machines Being Considered 209

16.3 The Choice of Equipment 217
16.4 Impact of the Operating System 219
16.5 The System Study 220

PART FOUR: THE COMMUNICATIONS WORLD

CHAPTER 17   EXPERTISE IN TELECOMMUNICATIONS 229
             17.1 Introduction 229
             17.2 Transmission Speed and Types of Terminals 230
             17.3 Data Controls and Security Factors 233
             17.4 The Use of Terminals 235
             17.5 The Use of Codes 238
             17.6 Facts Behind Standardization 242

CHAPTER 18   THE USE OF PROTOCOLS 248
             18.1 Introduction 248
             18.2 Protocol Functions 249
             18.3 Outlining Supported Services 252
             18.4 Procedures for Packet Switching 255
             18.5 Routing 258
             18.6 Flow Control Protocols 259

CHAPTER 19   KEY TOPICS IN DATA COMMUNICATIONS 263
             19.1 Introduction 263
             19.2 The Plain Old Telephone Service 266
             19.3 Speed, Bandwidth, and Switching Capabilities ... 268
             19.4 Modulation, Demodulation 271
             19.5 The Line Controller 276
             196. Sending and Receiving 280

CHAPTER 20   REALTIME, TIMESHARING,
             MULTIPLEXING AND LINKING THE PC
             TO THE MAINFRAME 283
             20.1 Introduction 283
             20.2 Batch and Remote Batch 285
             20.3 Realtime and Timesharing 286
             20.4 Multiplexing and Concentrating 290
             20.5 Frontend Processors 292
             20.6 Memory to Memory 294
             20.7 Connecting PC and Mainframes 297

CHAPTER 21 **GOOD NEWS AND BAD NEWS ON THE PC-TO-MAINFRAME LINK** **299**
21.1 Introduction 299
21.2 A Systems View of PC-to-Mainframe Communications 300
21.3 The Bad News: Logical and Physical Differences 304
21.4 The Information Center Strategy 307
21.5 Good News: The Solution of a System Architecture 309

CHAPTER 22 **USING THE NETWORK** **312**
22.1 Introduction 312
22.2 Routing, Monitoring, Journaling 313
22.3 Network Optimization and Utilization 317
22.4 Value Added Networks 320
22.5 Videotex, Electronic Mail, Home Banking 323

APPENDIX 1 .... **330**

APPENDIX 2 .... **335**

INDEX **338**

PREFACE

The application of microprocessors to business, to industry, and to our daily life brings a revolution in communications, financial systems, manufacturing, military equipment, and in technology itself. More precisely, it represents a knowledge revolution.

By the end of this decade, we will be a technology-driven economy. Why? Because the personal computer offers individual access to large databases at an affordable price. We live in an age when information transfer has become most important. Some say it represents 50 percent of the gross national product—and it is growing.

Such statistics affect all of us. Soon we will have pocket computers that students can bring to classrooms and can easily be used at home. The dramatic drop in the cost of computer electronics is changing the way businesses operate. Thus, we see the emergence of workstations supporting a data communications environment much more efficiently than earlier host-to-terminal disciplines.

Today, 95 percent of all American homes have telephones; however, data transfer is much more efficient than communicating by voice. With the greater variety of personal computers, the accelerating pace of technology, and changes in cost-effectiveness, new horizons have opened up. As a result, we have to rethink most of our opinions regarding teaching computers and telephony, and the ways both can be used together.

This text covers the use of personal computers, microcomputers and terminals, and the functions of data communications networks, and includes all information necessary for implementation. The presentation does not assume anything in the reader's background.

The text is divided into four parts. Part one stresses *a systems concept*. It introduces the personal workstation, provides perspectives in information systems, discusses how the office of the future may look, and outlines a specific situation where senior executives were introduced to personal computing.

Part Two provides the reader with an appreciation of the most valuable tools technology brings to our service. *A system is for the end-user*. This is just as true of office automation as of the personal computer itself.

Decision support systems can range from simple charting and spreadsheets to expert systems. In all cases, the purpose is to increase our capabilities through direct computer support. Part Two includes examples of distributed information in both a manufacturing and a merchandising environment.

Part Three presents some recent technological advances. A case in point is semiconductors. One chapter focuses on tomorrow: how the biochip may look;

are we going to move from personal computing to computational living matter? Interaction between man (the end user) and the information contained in the machine is also covered here. Not only are man-machine interfaces and voice input/output covered, but the medium in which information is stored and the way to judge the personal computer is also brought under focus.

Part Four addresses itself to the communications challenge. The first chapter in this section treats issues that represent a certain expertise in telecommunications. It explains protocols; includes key topics in data communications (from bandwidth to modems); introduces the concepts behind realtime, timesharing, and multiplexing; and ends with a discussion about the use of networks.

With new and interesting products appearing on the market, it is time that computers and communications people work more closely together to plan and implement the integrated information systems of the future. They will need much assistance from database specialists and end-user functions experts—but above all, they will need lots of imagination, hard work, and wisdom to train the user.

An efficient information system depends on more than software and hardware. If a system is to work, we have to consider the most important factor—*people*. Some of the challenging questions are: How do we enhance management's decision-making capabilities? How do we motivate our employees to achieve the highest level of productivity? How do we establish workable, cost-effective solutions? How do we select the software and hardware suited to our requirements? *The aim of this text is to help the reader answer these key questions.*

Let me close by expressing my thanks to everyone who contributed to make this book possible: my colleagues, for their advice; the organizations I visited in my research, for their insight; and Eva-Maria Binder, for the illustrations and the typing of the manuscript.

Valmer and Vitznau
Dimitris N. Chorafas

Part One
A SYSTEM AT WORK

Chapter 1

PERSONAL WORKSTATIONS

1.1 INTRODUCTION

The term personal workstation (WS) refers to one intelligent terminal per desk. There isn't room for two terminals on a desk. One must do all the work.

Personal computing may sound like a data processing (DP) orientation—but this is only part of the picture. The use of personal computers (PC) and word processing (WP) software makes the use of independent word processors unwise. Communications protocols running on the same PC do away with specialized terminals for electronic mail and a host of other applications.

We should capitalize on software to specialize the hardware functions. We should also capitalize on the fact that products are convergent. We must look at the PC as an integral part of the mainframe resources. The need for regular communication between PC and mainframes derives from this simple observation. Integrating these machines is a challenge and our goal should be to create a universal workstation serving a variety of functions. For example:

• secretaries need word processing capabilities, also archiving and communications

• investment bankers and brokers need instant market quotations

• managers require a variety of planning tools and pointers to corrective action

The same terminal can meet the needs of these. The same workstation (WS) can be used by managers, by professionals, and by supporting staff for various operations. The software makes this possible.

A whole infrastructure must be created to run this system in an able manner. Both users and developers should be aware that we are far from having reached the ultimate in technological evolution. The following ten developments in computers and communications will characterize the last fifteen years of this century:

1. Very large-scale integration with the 1 megabit (one million bits) chip becoming the standard for the 1980s (This advance was announced by Japanese companies in February 1984)

3

2. Universal, inexpensive broadband communications through satellites and optical fibers (we are already benefitting from this application)

3. Increasing use of voice recognition/voice answerback, leading to English language man-machine communications

4. Self-sustaining associative databases with relational characteristics

5. A new generation of very high-level 4th generation programming languages with database structures, precompilers, and system commands playing an increasingly vital role. For the first time in the 30 years of the commercial/industrial implementation of computing machinery, we will no longer need to tell the machine *what* and *how*. It will be sufficient to specify *what* we want done.

6. Optical disc memories making feasible electronic library solutions

7. Cryogenics or biochips equating biological densities

8. Hardware and software adaptable to environmental stimuli beyond the level of expert systems

9. Teachable computers, leading toward the use of brain-type metalanguage and the implementation of mechanized intuition

10. By the year 2000, we will most likely experience brain augmentation through machine-based high level intelligence.

Sounds far away? Let's only remember that 16 years ago man landed on the moon. It seems as if it happened only yesterday. Another 15 years is tomorrow.

1.2 DEFINING AND SETTING UP THE WORKSTATION

Personal computing is done at intelligent, programmable workstations. Workstations are both logical and physical at the same time. The *physical workstation*

- is a visual display unit
- contains a single board computer
- includes disc storage
- incorporates communications devices and other input/output (I/O) media

The *logical workstation*:

- is the address of all input
- provides local storage (microfiles) including text, data, image, eventually voice
- contains personal computing routines and
- executes input/output operations

By definition, the logical and physical characteristics represent the functions a workstation can support.

Database languages, spreadsheets, and packages for providing financial planning to producing color graphics have imposed an entirely new definition of what an intelligent workstation is and what it can do. The workstation has become much more than at-hand electronics. The whole issue of decision support rests on this reference. Executives often get where they are because of their analytical capabilities. However, on the job, they rarely use them because they lack the right tools. This is the primary subject to be considered when we talk about workstations.

An individual's analytical capability is improved through the use of personal computing, as is communications, databasing and the personalization of support facilities. Each of these functions imposes different demands on: information exchange, interaction rate, and internal processing. Our chief concern is to find system integrators and the software/hardware specialists with the appropriate know how. However, the field is booming, and it may be difficult to locate the best people for this job. Many go into business for themselves.

It is essential that the computer and communications specialists understand that the most important part of the system is the end-user. They must talk with the people who will actually be using their automated aggregate. They must get their opinions and get a feel for their attitudes. When any office automation system is implemented, we must be prepared to conduct a lengthy communictions and skill development program. It is also necessary to devote considerable effort and time to train people to use the system as it was designed to be used.

Efforts should center around the concept of the multifunctional workstation. Managerial and knowledge workers will simply not use systems where one function is carried out on one set of equipment and another on different units. From each individual workstation the system should be able to provide the following services:

- electronic mail
- various datacomm applications
- voice messaging
- teleconferencing
- decision support systems
- expert systems functions
- database access
- reminder service
- personal calendaring

User- rather than *technology-driven design* should be of main concern. Following a careful user-needs analysis should come the critically important user-training phase, followed by a pilot trial.

By improving analyst/programmer productivity an astonishing 3,000 percent,* Fourth Generation Languages make it feasible to proceed with prototyping. A prototype is not a theoretical system. It is a working model that can be implemented on the machine and for the environment for which it has been conceived. The prototype may include traditional data processing, word processing, document sorting, electronic mail, etc. It can be done through a spreadsheet or 4GL† database language, and it should be specific to the job—not a generality.

Prototyping can be done by the specialist or by the end-user. A senior executive said, "When it comes to my thoughts or ideas—I do it myself. When it is document handling, I ask my secretary to do it. When prototyping gets complex, I call the specialist."

At the end of the pilot trial, which should be made under realistic operating conditions, the host department must decide whether it wishes to take the final step of going fully operational with the system. The transition is then from prototyping to optimization, in an attempt to improve response time as well as other issues the pilot test may have brought to light.

This should be strictly a decision that the host department's management must make at the appropriate time, in addition to finding the needed funding to carry this decision out. The systems specialists should, however, continue to provide a monitoring and consultative function, including provision of an impact assessment or evaluation.

Prototyping and pilot trials are necessary to assure that professional workstations are being used in a manner that improves the efficiency by which data is collected, processed, and presented. However, it is necessary to be realistic. It is not an easy job to change bad practices of 30 years. It is necessary *not to program the devices by their functionality—but in terms of the functionality of the applications we expect from these devices.*

Figure 1.1 presents the logical component parts of a workstation. Applications programs should be retained on hard disc and loaded to the central memory when called by the user. They should then execute under a commodity operating system (OS) in the workstation. The operating system should be a broadly used commodity offering, preferably, CPM for 8-bit per word (BPW) engines; MS DOS for 16 BPW; Unix for 32 BPW. The concept of concurrent operation allows simultaneous input, storage, processing, and output at the user's workstation.

At the same time, a study should be made of the system integration. The first thing to realize in a workstation project is that the new equipment will both present new, never-before-available functionality and replace units already in existence.

During the last twenty years, office functions have been performed by a variety of machines that must now be integrated into the workstation: electronic type-

---

*Sounds incredible? Yet, it is true if we know how to go about it.

†4GL stands for Fourth Generation Language. The same family of programming tools is also called Very High Level Language (VHLL) in contrast to the now obsolete Cobol, Fortran, etc—originally known as high level languages.

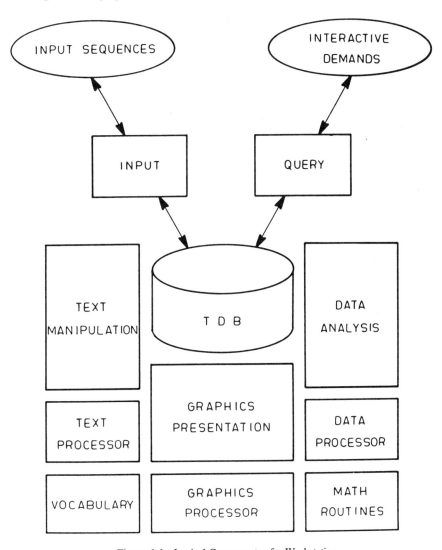

Figure 1.1 Logical Components of a Workstation.

writers, filing cabinets, microforms, calculators, timesharing terminals, telephone sets, telex equipment, copiers and telecopiers, facsimile, videotex units,
and stand-alone personal computers.

These are examples of the supports that, at different degrees of implementation,
have assisted in performing office functions and people have grown accustomed
to their use.

Therefore, when we design an Office Automation (OA) system we must
consider the user's images in terms of tools and their usage, his readiness to

change, the extent of training necessary, and steadiness of the semantics being used to ease both acceptance and training.

For all practical purposes, system integration takes place when the following twelve steps are linked together:

1. Architectural and conceptual

2. Functional, at the manager's and professional's level (on a multipurpose, intelligent workstation)

3. Training for a knowledgeable, homogeneous implementation

4. Procedural, including the study of all functions not yet automated

5. Operating systems, as the fundamental common ground

6. Integrated software, and other horizontal routines

7. Programming languages for vertical applications

8. Topological, including distributed databases, central resources, and public databases

9. Data communications, protocols and gateways: PC-to-mainframe local area networks (LAN), long haul, value added networks (VAN)

10. Channel integration—voice, text, data, graphics, image, with Private Branch Exchanges (PBX) as pivot point

11. Hardware at the WS level

12. Central computer resources (machinery for text/data warehouse, big communications switch)

This is a world of converging technologies. The real payoff in office automation does not come from automating individual tasks, but from linking a series of components to allow information to flow freely in many forms.

This is consistent with the definition of Office Automation which integrates personal computing, business planning through spreadsheets, business graphics, decision support software, expert systems, document handling, communicating with public databases, electronic message systems, voice mail, and the ability to network all office equipment to the mainframe.

1.3 SOFTWARE/HARDWARE TRENDS AND TECHNOLOGIES

An overview of trends and technologies in workstation development will necessarily involve the goals to be reached in terms of functionality and lead to a system definition. The latter needs to focus on objectives, reflect relevant industry milestones, and account for our evolving answer to managerial and knowledge worker requirements.

Systems solutions should reflect the fact that, as is currently projected, hardware costs will decrease an average of 32 percent while labor costs will increase an average of 10 percent per year for the period between 1985 to 1990. The able use of information technology calls for a careful look at such statistics.

We should keep in mind that the workstation facilities we now have available are the result of a long line of technical achievements. Four dates are outstanding. In the late 1960s, the minicomputer emerged and captured a sizable portion of the mainframe market. In the late 1970s, the 8 bits per word (BPW) PC took business away both from the dumb mainframe-based terminals and from minis. Then the pace accelerated and in 1981, the 16 BPW PC cut into the 8 BPW market, eventually killing the 8-bit micro. In 1984, this was repeated with the 32 BPW engines—whether as supermicro at the expense of the mini, or as an intelligent workstation for personal computing.

The year 1984 is significant from both a hardware and a software perspective. With the new generation of 32-bit single workstations introduced to the market at a cost of less than $3,000, a new classification is in order at the hardware end (Apple's Macintosh is the first of the new 32 BPW species, marketed at $2,500 having features previously available in machines selling for $10,000 or more).

Marketable microcomputers can be divided into classes. The low-end personal computer costs less than $1,000, selling millions of units per year and covering slightly more than half the market. The lower end personal computer typically has an 8 BPW processor and can nicely meet within a system integration perspective, the requirements of a clerical workstation. However, its real market is in home computing. Managers and professionals require more powerful engines such as in the middle-range personal computer, in the $1,000 to $3,000 level, and has estimated 25 percent of the market. This type of machine presently features a 16 BPW microprocessor, 256 Kilobyte (KB) or more central memory, floppy discs (of an increasingly larger storage capacity) and a communications discipline. Portables (lap computers) fall between middle and low-end.

The high-end of the personal computer market is in the $3,000 to $6,000 range with 32 BPW processors, half a megabyte (MB) of central memory, and a 10 MB hard disc. It has about 15 percent of the market. This type of machine can be used alone or networked, is highly functional, has a lot of user-friendly features, and addresses itself in an able manner to managerial and professional applications.

The supermicro computer features a more than $8,000 price tag, has a powerful microprocessor, one or more megabytes of central memory and several hundred megabytes (MB) disc storage. This is a machine that can be called the professional engine (PE). It runs a relational database management system (DBMS), acts as a rear-end (database) engine, and helps tie together other intelligent workstations.

In terms of software, no better reference can be given regarding successive generation than the spreadsheet leading into integrated software offerings. The first spreadsheet to be introduced to the market was Visicalc. Designed by Daniel Bricklin and marketed by Visicorp (formerly, Personal Software), Visicalc constitutes the beginning of a new generation of software. Lotus 1-2-3 is the best example of the next generation, Symphony of the latest. Tables 1.1 and 1.2 identify the functional differences.

TABLE 1.1 Generations of Spreadsheets and Integrated Software

"0" Generation:	Independent Packages (Spreadsheet, word processing, graphics, etc.)
"1/2" Generation:	Communicating Packages Can exchange files, but under user command
1st Generation:	Integrated Software File exchange is transparent to the user

Two Tendencies Prevail

"Bring Together" Existing Packages VisiOn EasyPlus	Really Integrated Approach 1-2-3 Supercalc 3*
2nd Generation	Further Integration Symphony CA Executive Intuit**

*Supercalc 3 is Supercalc 2 plus graphics and text editing.

**Intuit (by Noumenon) does away with the operating system, while supporting spreadsheet, word processing, database management, etc.

TABLE 1.2 First and Second Generation of Integrated Software

Supported Functions	1-2-3	Symphony
1. Database	Original Design	Improved Design
2. Spreadsheet	Original Design	Enlarged Design
3. Graphics	5 Graph Types	8 Graph Types
4. Word Processing	—	Yes (with Edit, Word Wrap, Insert/Delete, Erase, Copy, Move, Search and Display, Substrings, Headers)
5. Windows	—	Yes. Overlapping Options. Window on top is the active one.
6. Communications	—	Yes. File Transmission. Log-On to prestore, remote databases, suspend datacomm session, analyze data.

Other Integrated Software Offerings include functions like:

	By
7. Query	Vision
8. Form Generator	CA-Executive*
9. Tutor (Online Tutorial)	CA-Executive

Start/Stop and BSC Protocols become commonplace, and the DBMS facilities can be used as a fourth generation programming language.

*Computer Associates International

There is a difference between charting (which refers to simple bar, pie charts, and histograms, converting spreadsheets and tabular formats), and graphics (which includes icons, multiple windows, and editing of charts). In the word processing offering there is a need for a word processing style sheet to help edit on preferences (parameters) and, later on, to provide a tiling (stepping stone) pagination. Similarly, in integrated software, we would like to see calendar management identifying who has the right to call meetings, change calendars, authorize changes and update all calendars of communicating workstations.

The coming generation of integrated software will include the following:

1. Creation of system commands (shells) making all links (and other supported functions) fully transparent to the user

2. Encryption capability (first SW through password, ID; then SW/HW)

3. Integration links, word processing (WP) to electronic mail (EMail); calendar to EMail, etc.

4. Voice editing based on voice datatypes (digital encoding) and playback capabilities

5. Project management incorporating automatic reporting on plan/actual, update, Pert presentation

6. Budgetary control capability handling budgets, receiving financial reports, producing highlights, charts, and exception items

7. Expert systems implementing a knowledgebank (rules), presenting conversational reports, justifying the suggested course of action

The issue to concentrate on is added value. That is what the customer wants and the wizards of the software business are eager to offer it.

These are the tools that make it possible to increase managerial and professional productivity, if used correctly. For the executive and the knowledge worker, four main elements contribute to productivity. First is the ability to communicate information through electronic messaging. The amount of information we move around far outweighs what we produce. The automation of the communications function leads to significant productivity improvements. People working on the same project can exchange work elements through electronic mail, computer aided design/computer aided manufacturing (CAD/CAM), databasing/datacomm (DB/DC), and so on. It enables closer collaboration, better scheduling, and better quality.

The second element contributing to productivity is the integration on the same intelligent workstation of disperse functions spread on different mainframes and specialized dumb terminals. By having a personal computer at our desk we increase our access at the time we need it, and at the required level. Tuning becomes a matter of personal choice.

The third element is the facility to experiment for reasons of financial and operational planning, through easy-to-use tools such as spreadsheets. For instance, a leading bank found that complex investment proposals that took up to seven months for a professional to prepare could be produced in two hours through the use of a personal computer and spreadsheet. Personal software/ hardware tools increase our ability to access and manipulate the information sources we need—whether these sources are inside the company or outside.

Fourth, freedom from routine work permits high paid individuals to concentrate on knowledge production. Quite often managers have a distorted perception of the way in which they allocate their time. There is plenty of fallacy in management time distribution.

In the "knowledge society" in which we increasingly live, the trend is for manual labor to be replaced by cheaper and more efficient computers and robots. We must continue to produce knowledge; to do so we must continue to upgrade our skills, methods, and equipment. This course must be followed in order to remain competitive.

1.4   INTERFACING TO THE END USER

As Figure 1.2 demonstrates, the workstation concept is at the peak of a lower pyramid sustained by four technologies: dedicated microprocessors, local area networks, local databases, and gateways to mainframes. It needs appropriate software support to interface well with the end-user. Such interface will involve video presentation; touch screen or graphic tablet; the use of mouse, menu, and scroll; and eventually, voice input/output.

For man-information communication purposes, attention is being increasingly paid to prototyping and the use of an encapsulated object environment that is *icon-oriented** and able to support windowing on the workstation screen. There are several reasons for going in this direction. They are productivity, flexibility, interactivity, and reduced maintenance cost for software and data.

The new generation of systems will be easier to modify, and modification will be enhanced by the environment itself. This is fundamental in fulfilling the mission of transferring advanced technologies into future products; identifying, acquiring, and evaluating evoluting knowledge, assessing projected opportunities; and incorporating advanced concepts into new products.

As of 1984–1985 the identifiable end point systems are object-oriented, involve an end-user interface definition and are, by all counts, knowledge intensive. The design of man-information communication systems calls for the employment of metaphors, that is, ways of describing what the computer is doing as contrasted to what the people do. (A metaphor helps by alluding to something else, sending a message, providing an interface with common technology.)

---

*Icons are familiar objects which we present on a video screen for explanation and/or use in programming.

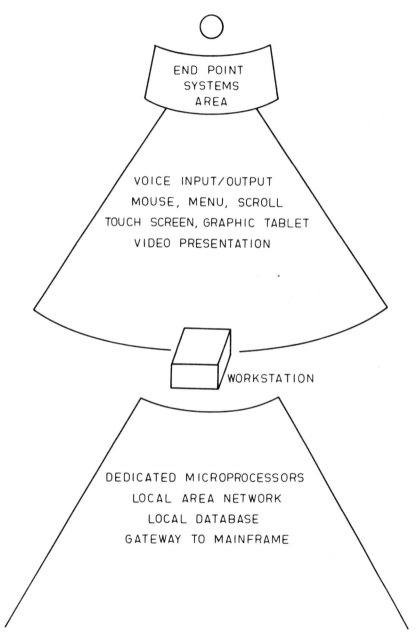

Figure 1.2 From Systems Components to Systems Integration.

The object of designing for the end user systems area is integration, and the link to the further services needed to perform one's own functions and contribute to those of others.

Projecting a workstation is both a local and a system activity involving the divisional assignments between the workstation, the local database, the files in the database, semaphores to files, memo-posting procedures, electronic mail capabilities, and local area networking (LAN).

Communicating workstations will increasingly become the rule. Industry specialists project a widespread dependence on integrated workstation networks using LAN. In an office automation environment, the workstation at each layer can function well through a workstation connected on a baseband, however, internetwork communications requires broadband support.

On a long haul (wide area) basis, communications links must assure: connection between the local databases and the company's mainframe(s), access to public databases, consistency in local update, and global text/data consistency. Both local and long haul communications should account for the growing importance of management graphics.

Extensive use of computer graphics (including color) and other advanced end-user interfaces such as voice recognition/synthesis is also projected for the mid-1980s. Voice recognition/synthesis is significant as developments point to the high use of voice (at least for input), the increasing implementation of store and forward (voice, text, data, image), voice memos, annotations, and the able handling of help messages.

Human engineering aspects will increasingly dominate design approaches. The semantics of a good end-user interface must be consistent with the semantics of the work being done.

The syntax of the interface should be as natural as possible, appearing intuitively obvious to the user. Accomplishing the same function by multiple variant forms must by minimized, and the use of variant forms for similar functions eliminated.

The system must offer selectable levels of aid and guidance for users of different degrees of expertise. The end-user language needs to be easy to comprehend, use, and remember. At the same time, the programming language must be powerful; each user entry should perform lots of work, and the language should also be expandable, accepting overstructures. Examples of system flexibility are the following editing functions being supported:

- Cursor Up, Down, Right, Left
- Clear Screen
- Erase to end of Field
- Character Insert Mode On/Off
- Delete Character in Field/Line
- New Line

- Forward/Backward Field Tab
- *Home* Function

An indent function is helpful in moving the cursor one tab stop to the right of its current position. Attribute highlighting can be extended into a significant range of facilities: skill/nonskip, swap of field tab, suppress pacing, bell ringing, and controllable indicators.

Testing and evaluating are the only means to ascertain the functionality a system presents. This is particularly true of the media directly visible by the user such as cursor positioning, menus, windows, and document creation facilities.

Track balls have become popular devices for cursor-positioning purposes. A track ball is generally known as the *mouse*. Moving the mouse causes the cursor on the screen to move directly to the desired location. This is a simpler approach than pressing keys on a keyboard. However, the use of a mouse as a pointing device is no cure-all solution. The mouse is useful for spreadsheets and anything relating to menus and windows. It is not recommended for heavy-duty word processing applications. Here, solutions must be provided that confer enough technical benefits upon their users.

Windows are independent format handling devices that respond to standard input/output calls. They also answer commands to manipulate attributes including location, size, font-usage, exposure, and keyboard status. Typically, a window is associated with the raster-scan video screen upon which it is displayed. Windows may or may not have a one-to-one relationship with processes.

Every window has an independent multifont map. It attaches itself to a process-group to which a signal is sent whenever the window changes size or location, becomes exposed or covered, gains or loses current keyboard window status, etc. Only the user has control over the windows in the system, but a window can be manipulated by a process in its process group, or by a process with write permission on the window.

All windows, upon creation, are entered in the file system as character-special devices. In the directory, they have a filename specified by their user-supplied label. Windows are dynamic in that their location, size, exposure, and fontmap can be modified at any time under user/program control. When a window is created, it can be used in a variety of ways by the process that created it.

When a window is covered, output to it can be saved in a buffer and displayed when the window becomes exposed again. This way a process will not be halted because it wants to output to a covered window.

Windows lead to the implementation of a simulated virtual memory environment allowing combinations of applications to run concurrently which could never fit into ordinary personal computers. User creation and control of windows is accomplished through a set of calls. These allow a user-process to:

- make and initialize a new window
- draw or erase a window

- insert a selected window
- obtain the current state of a window
- modify the current state of a selected window
- manipulate the fonts utilized by a selected window
- obtain the current state of the display to which a given window belongs
- switch keyboard input to a selected window

Windows may occupy independent and changeable rectangles of videoscreen surface and may overlap each other. They help users move easily between spreadsheet, word processing, graphics, and various other packages.

Document creation and filing is software-supported and enhances the ability to tear off stationery to create a document by opening a folder, duplicating, inserting document on folder, filing away, dropping the stationery into a waste basket (tear off), and by monitoring the printer if a hard copy is desired.

There are two ways to document handling integration: have one applications package do everything; or move text, data, and graphics from one application to another in a transparent manner. Both ways lead to the ability to provide results by merging text, data, and graphics.

1.5 CONNECTING PERSONAL COMPUTERS AND MAINFRAMES

We have said that the stand-alone workstation is of relatively limited value. The future depends on the ability to communicate, in a way transparent to the user, from any workstation to any other workstation, as well as to and from central resources.

With networking, the following functions are critical:

1. It should be possible to provide both real time and deferred (store-and-forward) facilities, with little distinction between the two.

2. The system should support communications without distinction as to text, graphics, and voice.

3. The media supported by the system should combine in any way desired by the user.

For instance, it should be possible to embed voice annotations in text, or graphic images at specific points in a voice message. It must also be feasible to integrate additional, task-specific media required by the user. There are different reasons why this is necessary. First, the satisfaction of the growing ranks of corporate workstation users. These users consider it critical to tie into the mainframe. The second reason regards the reduction of processing loads on the mainframe. Third, with intelligent workstations, the processing job is brought where it belongs—

to the workbench. The fourth reason is the tailoring of the application itself to the individual needs of the user. The fifth is the benefit in considerably reducing response time. This is achieved because computer power is now dedicated to one job, and, therefore, has the ability to do it well.

Without doubt, the simplest approach for providing the logical personal computer-to-mainframe link is to have the workstation emulate a nonintelligent computer terminal. This is the way that most computer users interact with commercial databases.

Such a timesharing mode requires: a terminal emulation software on the workstation, a modem, and a communications line. This is presently the most common form, but not a very effective one. It allows the personal computer to look at the data, not to store them and manipulate them.

One step up is the ability to download data. This permits the end-user to call information from the mainframe and store it. The data is in raw form, and software is necessary to get it from storage into the spreadsheet.

An improvement over simple downloading is to add software to allow the data to be formatted correctly. Thus, data can go directly from the mainframe to the workstation into the proper slots in the desired program. Whether downloading or upline dumping, this type of communication requires handshaking protocols at both ends.

A more sophisticated approach is to have the microcomputer and the mainframe run the same programs so that translation of data by the personal computer is not necessary. Process-to-process communications, like all other modes, can also take place between workstations. In recognition of the importance of enhancing communications capabilities, IBM introduced two desktop machines designed for such link: the XT/370, and the 3270 PC, which doubles as a computer and as a 3270 family terminal.

The 3270 PC holds four datacomm sessions and two scratch pad sessions. The XT/370 operates under VM. Other devices can also communicate with an IBM mainframe. If alien equipment supports the RS 232C, then the connection is only at the physical level. If a 3274 concentrator is used, it includes data link and teleprocessing capabilities. Or, if the concentrator is 3279, the system supports raster graphics. It can also be a loop connection (8100, 4700). A Peer-to-Peer long haul connection can be supported through start/stop, BSC, SDLC, X.25, and X.25 with cryptography.

Independent vendors are presently offering dozens of products to link personal computers to big computers. They are both interactive, with process-to-process orientation, and file-to-file for bulk transfer. At the same time, increasing attention is being paid to encryption capabilities (passwords, authorization, authentication). It also implies that the document will be scrambled, and it will no longer be possible to read it without the right code. These are issues to be studied prior to instituting the new computers and communications aggregate, not after. Encryption capabilities will be particularly important in long haul communications for instance, between, the central computer resources and those in the branches, sales offices, factories—or those installed at customer organizations.

Such developments take place at the following three levels:

1. personal (at the base of the information pyramid)

2. workshop/departmental

3. corporate (top of the organizational pyramid)

As the last three years help document, with on-line terminals (personal computer or no personal computer), most business people don't want to compute—*they want to communicate.*

Chapter 2

THE OFFICE OF THE FUTURE

2.1 INTRODUCTION

The office should be viewed as an intelligent machine, an extension of the human mind. This helps to better structure facilities and functions, becoming the general framework within which systems design will be made.

The second basic reference is the level of office automation (OA) we are talking about. Five years ago, even three years ago, the most emphasis would have been placed on clerical and secretarial work. This is no longer the top field. Today, fully integrated office systems are designed for the management and knowledge worker (professional) segments of the office population. That is where the real growth area lies for office automation systems in this decade. It is also the rewarding one.

Studies show that managerial and professional staff account for more than 70 percent of office cost, yet until recently, this segment has been untouched by system studies. Reasons for this are as follows:

1. Early OA systems were not particularly user friendly.

2. Busy executives were unwilling to take time to develop the skills necessary to effectively use them.

3. Early systems lacked the capability of making available in one compact device all the functions managers and professionals perform.

4. It is always difficult to measure mental productivity.

5. Only recently has continuing education been accepted as the key to professional survival.

The latter reference underscores the fact that a major problem in automating the office is the human element. The finest system can be an expensive dust collector if people are unwilling to use it.

The importance of in-depth user training at all levels of implementation cannot be overemphasized. And, it is vital that users do not get the impression that the introduction of office automation systems will result in a need for employees with a diminished skill level. To the contrary, the automation of office tasks will free users from routine work and make it possible to carry out more challenging and interesting activities. It is important for users to be brought into the planning stages as early as possible and be made to feel part of the whole decision-making process. With proper training and a sense of participation, user enthusiasm can be raised most significantly. This attitude and environment has to be created if office automation projects are to be successful.

Finally, an office automation study must be thorough. No shortcuts should be allowed in the planning process. It is also necessary to have the receptiveness and full support of senior management even if it proves time consuming. Furthermore, the extent of planning required for both technological and organizational aspects should not be underestimated.

2.2 DIRECTION IN OFFICE SYSTEMS DEVELOPMENT

The office automation system is evolutionary. There is not going to be a day when an office system will be described ''ready'' and ''perfect.'' Improvements are continuously made.

Office systems technology both leads and follows customer implementation. Hence, there is a need to build a system architecture. An office systems architecture provides for continuity. From continuity comes the notion that we can turn the workstation into a window to the environment of the organization.

It is just as important to realize that the value of an office system goes up as the number of users increases. Unless we plan properly and proceed efficiently, it can take several years to reach a critical mass.

We must also appreciate the need for managing information as a product. This is no self-evident goal, but experience demonstrates that in most financial and industrial organizations, information, for the most part, is not managed, is available in overabundance or not at all, is seldom timely and complete, and is provided at an undetermined cost.

This is the result of having today's approach to information management based on yesterday's concepts and technologies. However, it must be realized that times have changed.

Every six months something significant happens that alters the way we look at the work place. According to our best projections, this will continue to happen over the next ten years.

While change cannot be stopped, it can be managed by planning ahead to achieve reasonable goals. This is a new approach altogether: Over a good 30 years of computer implementation, users reacted rather than prevented; and they discovered ''after the fact'' instead of planning the application.

Numerous examples make it evident that the concept of avoiding problems by avoiding change leads logically to an even more absurd conclusion: avoiding the solution of problems by avoiding system use. The cost is there but the results come rarely, to everybody's consternation and concern.

The time has come, therefore, to take a different approach. Invariably, this will involve a major role for microsystems in banking, business, and industry. It is expected that the number of micros in offices will soon exceed the number of stand-alone word processors. Still more significant, personal computers and local area networks are beginning to be seen as a replacement for mainframes with many associated terminals.

The 1982 National Computer Conference in Houston displayed a variety of both the new and the familiar. However, there were no mainframes on the convention floor, while several of the small systems on display rivaled older mainframes in capability. And that was in 1982.

Interactive facilities have become important as they support and promote visual thinking. Graphical input is in its early stage of development, but graphical output leads the way toward fully supported means of dynamic presentation. Once we appreciate the principles of interactive displays, noninteractive ones become trivial.

The cathode ray tube (CRT, video, softcopy device), which was invented more than fifty years ago, is likely to dominate the display market for decades, especially where graphics, color, or both, are needed. Pictures are a natural, venerable, and most effective way of communicating. They can greatly enhance a dialogue between the individual and information.

Through video presentation, text can be shown in ways that are impossible on paper: Parts of the screen can be dynamically overwritten, can be highlighted by blinking, exceptions can be presented instantaneously in different colors, three-dimensional objects can be drawn, and the viewing position can be changed by rotating the object.

Output media are undergoing a silent revolution. Many financial institutions and industrial concerns today expect to see far more graphics used in offices for enhanced communications, even as much as half of all office communications in graphic form. Also, there is considerable interest being expressed in voice mail systems for increased efficiency.

Never think that to see the future we should talk to vendors. Vendors have vested interests in the past, and this makes them look backwards. The way to appreciate the future and benefit from technology is to talk to the pioneer users because they are the only ones who know what's coming. They can help us avoid redundant effort, and they can give advice about avoiding the mistakes that have already been made.

An example of an informed, progressive user is the United Services Automobile Association (USAA), a large property, automobile, casualty, and life insurance company with headquarters in San Antonio, Texas. Since the 1950s, USAA had been an IBM mainframe user. In 1979, management decided to implement personal computers (PC) and local area networks (LAN), thus making this insurance firm one of the first users of intelligent workstations. For LAN,

USAA chose Datapoint's attached resource computer (ARC) system. The network consisted of a variable number of workstations, where each station was able to communicate with the others, and with shared resources (disc, storage, printers).

The new claims processing system for this configuration was developed over a ten-month period by USAA. Then a six-month test followed, using one ARC system with fifty workstations attached. Management was satisfied with the results, so another thirteen ARC with 750 workstations were installed, covering almost all claims processing. As claims calls or letters were received, the claims handler took the information over the phone (or from a letter) and entered it directly on the workstation.

If the caller gave the policy number, the policy information would be immediately retrieved from an IBM mainframe (via Datapoint's 3270 emulation) and moved to the ARC's disc storage. If the caller did not know the policy number, a computer lookup would be needed. The claims handler then entered the rest of the information about the claim, making explanatory notes as desired.

The result has been that claims handlers now have almost paper-free desks and the system proved to be a significant success in terms of meeting the goal of processing more than 90 percent of new claims without paper claim file, and by improving the productivity of the claims handlers themselves. User acceptance was without reservation. Users asked for further additions and enhancements to the system.

The company's system planning aimed at providing computing power at the work groups and individual workstation levels. This has been what the user population applauded the most. Such developments do not come about on their own. They need a great deal of preparation, the right level of skill, and, above all, proper training.

An excellent example of this is the computer literacy program run in parallel to the introduction of professional workstations at the Mellon Bank in Pittsburgh, Pennsylvania. In 1983, Mellon Bank, America's twelfth largest bank in terms of assets (deposits of about $16 billion) began a program to make all of its employees, from the newest clerk to the chairman of the board, computer literate. Training included instruction in the use of personal computers and inquiry/ response languages.

Intelligent professional workstations are being installed in the offices of the bank's middle and upper management to be completed by the end of 1984. However, until recently, the Mellon Bank had a mainframer orientation.

Mellon Bank management expected to spend in excess of $40 million for these new tools oriented to the senior executives, managers, and professionals. Subsequent orders for units at its more than 300 correspondent banks could bring the total value of the order to more than $100 million.

The objective is to increase the number of people in the bank who use computers in their day-to-day activities from 30 percent (mostly branch tellers) to 70 percent within two years.

2.3 BACKGROUND FOR THE NEW TECHNOLOGIES

When we talk about employing the new computers and communications technologies, our first and foremost concern should be for the end-user. This is
feasible with intelligent workstations: The application drives the development
of the system.

This application orientation brings to the foreground the need for system
integration (which we will discuss in greater detail in the following chapter). It
also underscores the three areas on which a successful computer and communications system must react (see Figure 2.1). The most important components
are the system owner and the computer users. Next in importance are the system
architects and the consultants who bring in new knowhow. Down the line but
still vital are the vendors, both for the software and for the hardware.

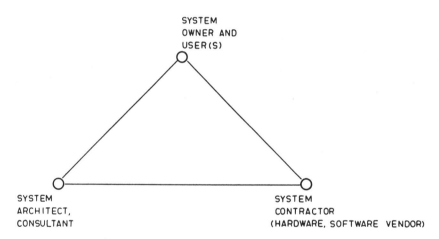

Figure 2.1   A three-party collaboration in building an efficient system.

An excellent example of office information systems strategy is given by Citibank in London. Here, the key issue in implementing a local area network and
personal computer policy was systems acceptance. As senior executives related
it, the very successful 1981 personal computer and LAN implementation in Forex
was the fourth attempt to get dealers to use a computer system. The first attempt,
which took place around 1976, and two more efforts in subsequent years, were
not very successful because the user was not the key element of the system.

While mainframes are two degrees removed from the user, as are minicomputers, communicating personal computers make it possible to readily identify
where and how we can use the information which we generate. Furthermore,

we can do so at the work place level, before centralizing the information elements or sending them to other functions.

This concept is so vital that it's wise to restate it. Communicating microcomputers now offer the opportunity to do exactly the opposite of what had been done. The process of first centralizing and then disseminating the text and data has been reversed to make it much more efficient.

The Forex application at the London operations of Citibank is a typical example of how big companies move into fully distributed computer environments. In late 1980, management became aware of the personal computer's potential and realized how cheap it would be to implement a user friendly solution, how potentially easy it would be to program and how easy its use could be justified at the work place. Once the first personal computer was bought and put to work (by a dealer who wrote the software to do what he wanted and demonstrated the results to his colleagues), the other dealers appreciated for the first time what could be done in their daily work by computers. Soon they realized that many functions could be done on the machine.

The Executive Vice President, attending a technical course, saw an Apple computer and said to his immediate assistant, "Take a look. It's very interesting." This changed the whole organizational mentality, and as the latter was to comment, "I am not at all a computer expert. My expertise is limited to spreadsheets."

While a basic reason for implementing LAN is financial, the rationale of effectively sharing text and data should always be kept in perspective. Many intelligent workstation implementations started from the financial operating position and from there, moved into managerial and operating effectiveness.

But in order to do something really worthwhile, which alters past, largely obsolete images and practices, we must have the support of top management. We must also be ingenious in cutting the red tape.

Lulled into complacency by the belief that computer users are locked into expensive systems they have operated for three decades, many computer executives have grown increasingly insensitive to the need for implementing successive stages in development. This has led to growing user disillusionment with mainframes. It is, therefore, not surprising that well-informed users became interested in the micro. Apart from being able to see, feel, and better control the application, the micro presently is the most efficient means for bypassing the big bureaucracy built around the mainframe.

The difference between using the intelligent workstation and the mainframe can be briefly summed up. It is the basic concept that has changed. We are finally moving out of the classical "EDP era" of the 1950s, toward mastering the computer and putting it to use.

The "EDP Connection" has its roots in the view that the computer is a glorified, electronic accounting machine. We know this is untrue; and yet for nearly 30 years few people appreciated the difference. The real power of computers and communications is the services they can directly provide to the end user.

Ten years ago it was thought the mini could bring computer power to the end user, but it was unable to because of the deep rooted "central processing syndrome" described above where the same old bureaucracy oversaw the mini.

It is therefore wise this time around to plan carefully. Nobody should be allowed to bet on the "status quo," even if there are huge investments in software that run solely on mainframes. And no organization should be shy about implementing personal computers and LAN because its information scientists failed to update their skills.

At the user level, the changes must be structural. Until a few years ago in all cases—and still the case in a large majority of installations—the way things function is that all development is being handled by the central computer resources. This creates bottlenecks, leads to misunderstanding about real user needs, tends to produce delays, and makes many computer projects too costly. User goals are rarely achieved. Although the main objective may be to help the end-user, there are also many other goals. Typically, these goals may be book-keeping, production of government reports, reports to management, and ledger of assets and liabilities. Yet the final purpose should be to provide assistance to the end-user.

Computer "power" must support the human elements rather than the big organizational red tape. Project teams must be agile, lean, and small. For instance, Citibank in London put a project manager and three analysts together to implement the Forex operation. The manager himself remarked, "Probably the team should be much smaller. With small groups it is a lot faster to do things." This is true for a number of reasons. First, the mainframe approach kept various functions separate. However, with intelligent workstations, all functions can be done at the workbench level, producing faster results. Second, the end-user must get personally involved with the application of computers and communications—from analysis to programming. We do have today at our disposal Fourth Generation Languages that are very easy for the end-user to use. Spreadsheets like Lotus 1-2-3, VisiOn, Multiplan, MBA, and CA-Executive, are some examples. It takes a few hours to learn how to work with them, and the end-user can then apply them to a whole range of requirements.

Another Fourth Generation programming tool friendly to the end-user is the database management language, particularly the query-oriented supports. IBM's SQL is one example. QBF (query by form) by Ingres is another. Experienced end-users would not find it difficult to program in BASIC, C, or Pascal.

In the Forex case I mentioned earlier, the applications software was originally written in BASIC by one of the dealers. The other dealers' reaction was positive; the problem was to have the micros accepted by the controllers.

Significantly, in this as in many other cases, the users themselves became interested in reaching an understanding on how to work in a workstation. A turning point in obtaining the controllers' approval was the definition of a single document to come from the processing operation and the way in which it should be designed.

Information scientists, end-users, and controllers came to the conclusion that there was no point having a lot of paper moving around. Quite interestingly, the same principle applied to the system analysis itself. The relevant documentation was produced after the system became operational.

Let's make no mistake in this example. Written systems analyses are necessary, but with intelligent workstations they are less detailed and the whole process moves faster. The users themselves should get involved in system design. In the Forex case, the result was the generation of a position-keeping system that handles multiple currencies, directly through point-of-origin data capture, dedicates very little to mainframe support, and is designed to simplify and make input/output efficient.

Another way of looking at this is that the people who generate the transactions are also the input operators. Management felt that this was obviously a much better solution because if the dealers use the personal computer network to input their operations they could have more reliable information.

As a management principle, this did not come around for the first time with the personal computer and local area networks. Management tried for years to apply new working principles, but dealers wouldn't do so unless they got something in return. What they wanted were results, and this the mainframe stubbornly would not give them.

2.4   NEW APPLICATIONS

Because of the versatility of personal computers and local networks there is practically no limit in the generalization of functions to be run through local intelligence. Applications classically handled by mainframes can be sucessfully converted to workstations.

The general ledger is an example. Both general accounting and cost accounting can be kept in realtime through personal computers. In fact, among the best early applications packages for personal computers has been Visicalc for financial planning. It is no longer the smaller firm that make use of such solutions. Reuters has replaced its mainframe timesharing by interconnected intelligent workstations.

The choice of applications depends on where an organization already stands in terms of computer usage and on its goals. Ironically, it is those firms that are further behind in office automation that have the greatest desire to embrace the new computers and communications technologies.

With the reviving emphasis on management information systems (MIS), decision support is a good field. Because experience is still thin in terms of cost effectiveness, a leading user commented, ''You may not cost/justify at the start. You take a chance and you see how it works.'' The chances, this same source observed, are very good because the system is close to the user, the user judges how to go, and which applications aspects to emphasize. For the same reason, intelligent workstations are well suited for control purposes: confirmations, consolidation, and checks on operations as they occur. This is done on-line to the database, while with the old mainframe and mini based systems, it was only possible to initiate transaction, leaving to subsequent phases the attention to database tuning or pruning.

Microprocessor-based media are further enhancing the interactivity of the workplace. With the graphic tablet, for instance, it is easier to complete trans-actions at the dealer's level, on-line to the local database, than write out the classical deal ticket. This proves to be very important to the end-user.

A trucking company evaluated three different ways of automating its dispatch-ing office (see Figure 2.2). The classical approach (the one that had been em-ployed for years) was a Service Bureau, third party data processing system. One alternative was to hire and use on the premises a small business system (SBS). These two methods, management found, had weaknesses, such as slow response time and low reliability. After all, one microprocessor had to do the whole job. The cost of either of these two methods was too high for the less-than-perfect results.

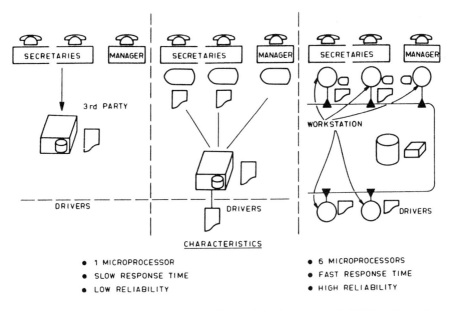

Figure 2.2 Comparing the services supported through local area network (LAN) to those of a small business system.

A solution to the SBS was to use the PC and LAN, taking advantage of the cost effectiveness that high technology permits. After a careful evaluation, man-agement chose this solution.

However, it is most important to start with good applications for local net-works. Typically, these will reflect truly distributed systems controlled by the users. A LAN is at its best where emphasis is on access to local resources with a user friendly interaction. A LAN will perform poorly if one is trying to use personal computers as just another mini, midi, or mainframe. Multiplexing the users is a poor approach; limitations in the power of the workstation will show up.

From supercomputers to distributed microcomputers there is a wide range of applications possibilities. While some 20 personal computers linked through a LAN and accessing a database can match the power of a small mainframe in terms of processing capacity, it is wrong to tackle the subject this way. Strategy and structure must use a system's competitive advantages.

The level of interactivity supported at the user's desk brings up advantages which, for the most part, are tangible. Theoretically, graphics presentations can be done through nonintelligent terminals hooked to mainframes or minis; but how many organizations are doing this? On the contrary, workstation-based graphics capabilities are becoming widespread. It is not only the personal computer that nicely supports management graphics; there is now on the market an impressive list of microcomputer software doing just that.

Integrated software offerings go from spreadsheets to graphics conversion, include word processing (WP), calendaring services, database management capabilities, and communications protocols. Many of these packages also feature different levels of computer graphics.

Let's return to the fundamentals. Computer graphics is the computer *synthesis* of pictures, generating charts and icons with the aid of a digital computer. In contrast, the *analysis* of natural pictures is called computer image processing.

Although the two activities have much in common, they are different disciplines and should be treated as such. Because of high resolution graphics and visual thinking concepts, a three-way division is developing charts, complex graphics, and icons. Charts are of the bar, line, and pie type. They represent the more classical approach to management graphics which, in a PC environment, is spreadsheet oriented.

Videotex, for example, is a graphics presentation means. For bar and line charts, alphamosaics pose no problems. In fact, charting can have different levels of sophistication in presentation. It can also support color. Smart people who use color charts standardize the color.

More complex computer graphics calls for six basic components:

- greater computer power
- appropriate software
- proper resolution in output equipment
- needed input data
- able user interfaces
- a training program able to change user images

Icons could be equated with complex graphics. I take them as a separate class for two reasons: first, because icon shapes can be used as standard modules/

visual thinking, and second, because the able handling of icon interfaces calls for a query language.

Under current practices, the presentation of different sets of icons can vary widely. The challenge is how the population of users associates semantics to them. The best combination is the simple icon. The worst are sophisticated designs (because of many errors), and those including literary description (slow reaction by user).

Graphic tests are necessary to measure the effectiveness of icons. Graphics concepts are needed to effectively deal with the new tools available in the modern information systems environment.

We have been talking of visual thinking. A related issue is visual programming. With visual programming, "what we see is what we get." For example, the user defines a screen and associates it to an expression without having to worry about file handling and structure. This is done through a high level interface. Processing is accomplished by generators.

These are examples of facilities that greatly enhance management productivity. They make information more comprehensive, they provide clear decision-making bases, and bring to the attention of management only what is important. Decision support, graphics presentation, color enrichment, and visual programming require intelligent workstations—and this underscores the importance of distributed systems.

Distributed computer power runs contrary to the old idea of trying to handle everything by squeezing it on one central system. There exist degrees of freedom not feasible until a few years ago; an example is the way to handle the database.

With intelligent workstations, there is no reason why we should integrate the organizational database on a wide basis. While this has been necessary with centralized and partly distributed systems, we can nicely handle logical files without a gigantic streamlining routine. This keeps everything small. The individual user can understand the system better. The personal computing, text/ database and data communications specialist can easily see how to manipulate any new function. Also, it becomes possible to always send information to the recipients in the way they want to receive it.

This makes it necessary for the gateway to restructure the information; but at each local workstation level, text and data can be linear, lean, and simple, and must be kept that way. For the first time, this makes feasible a double approach. Locally we can implement precisely what the user wants (microfiles). Centrally we must streamline, but this will be the exception, needed for the interconnection of the information elements (IE) and their consolidation.

Quite evidently, there are challenges. The most difficult part is not the hardware or classical software chores, but the way to rationalization. Even though the interactive pages (and the files behind them) are all personalized, we still must have a precise strategy about where to go.

Because the earlier personal computing applications have been standalone, one of the challenges is migrating from standalone to networking. For this purpose, we must capitalize on the strengths of the personal computer, more particularly, its flexibility. The system software is so versatile that there is no reason driving us to change with new vendor OS releases.

Another one of the challenges is extending the basic routines supported by the personal computer manufacturers and software houses. An example is provided by file handling. Reliability factors preclude keeping only one copy of the local files on only one piece of equipment, no matter how reliable it may be. Backup is necessary; and that's the mainframe's new role—a central text and data warehouse, and a big communications switch.

Systems problems, in other terms, have to be solved by the right applications studies. Solutions must be ingeniously worked out. Stereotypes will not help.

Chapter 3

NEW KNOW-HOW

3.1 INTRODUCTION

Information systems are now entering a new era following years of rapid technological change. This can be seen by the new technology used in microcomputers, in data and text communications; in local and remote databases; in the implementation of multimedia networks that handle data, text, voice, and images; and in the wide dispersal of multifunction intelligent workstations that interface with data and text file processors. The new orientation stresses office automation, text and databases, on-line interactive services (including journaling and security), new approaches to voice communications, and robotics for the factory. In the data processing environment, the current trend to go from batch applications to on-line will accelerate. However, a total conversion will take years to implement for the following reasons:

1. large software investments made with batch programs
2. the multitude of prerequisites implied by online solutions (see Figure 3.1)
3. the lack of system specialists able to fulfill such prerequisites

As an integral part of the coming evolution, networks will support text and data image solutions, voice store and forward, and a text and data entry facility within the broader concept of office automation. Network architecture will employ packet switching protocols which are applications independent and have become an international standard, and network control centers, able to do on-line diagnostics, keep quality histories, and provide assistance for self-maintenance.

System design will stress databases that will be integrated, thus making possible their partition and distribution in a layered, expandable approach; that are able to handle text and data (being increasingly supported by dictionaries); and that are protected through security mechanisms for authorization and authentication.

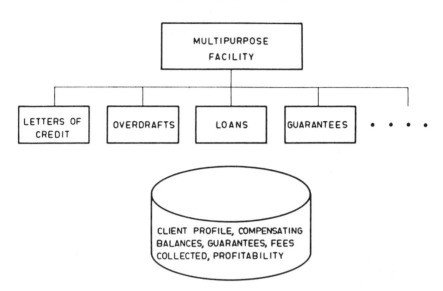

Figure 3.1   Interactive Client Image.

Such developments call for a new perspective and a different orientation than we have followed for the past three decades. Computers are no longer a primarily DP-oriented engine. They are vital parts of a broader communications system linking the user and his or her problem to the information through a secure and updated access mechanism. This must be done interactively, in a user-friendly manner, where no excessive training is necessary for the user, and by applying known technologies in an imaginative way.

3.2   TRENDS IN INFORMATION TECHNOLOGY

For changes to be effective, we must alter the image of information systems capabilities held among users (both management and day-to-day operators) and among the specialists. This involves continual training. This training should both increase individual know-how and update available skills in terms of the major trends in high technology and its implementation. For 1985, this leads to a confirmation of the application of personal computers and local area networks in financial and industrial environments which, in turn, underlines strategic perspectives. In the banking industry, for example, the spotlight focuses on home banking and electronic tellers, going well beyond the Automatic Teller Machine (ATM) level.

In the mid 1980s, home banking reaches its first level of maturity, broadly dividing into three basic solutions: Videotex; on/off intelligence through PC; and an intermediate stage. Similar references can be made in other industrial sectors.

The new range of applications is highlighted by a steady evolution and improvement in LAN architectures. The second LAN generation was introduced in early 1983 while the second generation of personal computers was introduced to the market in mid-1982.

Top management is one of the levels where PC and LAN are presently used to advantage and there are many applications already being used to replace minicomputers. Within this context, mainframes are phased out in terms of processing, but, as previously stated, they remain as a text and data warehouse and as a central switch. The datacomm nodes and microfiles are also distributed, with personal computers increasingly able to run large discs.

However, an evaluation of future trends cannot be based on what can be seen in the short range. Long-range planning is a prerequisite for forecasting developments in computers and communications during the remainder of this century.

Starting from the premise that hardware is mindless but can acquire an intellectual focus through software, successive milestones to the Year 2000 are leading to intelligent machines.

An example is with decision support systems (DSS) and expert systems. They both use facts, rules, and reasoning methods. But DSS were focusing on facts, had few rules, and only one method.

To reason, postulate, and learn, expert systems need knowledgebanks with hundreds of rules—and they are rich in methodology. The latter will be further strengthened as the expert systems become increasingly able to provide evidence, calculate uncertainty, reach conclusions and document them.

There is a parallel in terms of biological evolution: the first advance in information occurred in the carboniferous period when, for the first time in history, the world had more information in its brains than its genes. The second giant step took place some five or six thousand years ago with the invention of different forms of written communication—from hieroglyphics, to the numbering system, to Linear B. This allowed information to be stored extrasomatically.

Once the methodology had been achieved, the means followed. Gutenberg's invention of moveable type marked a qualitative development of this trend. The development of microprocessor-based intelligence may be viewed as the third quantum jump. For the first time in human history, extrasomatic storage of intelligence became possible. This is definitely a major change. The acceleration in the sequence of these three stages (the first of which did not involve the homo sapiens) is also impressive—hundreds of millions of years, thousands of years, and finally, decades. Within this much broader perspective, the personal computer is no more than a transience, and an important one for that matter. For the first time, we are able to associate data processing (DP), databasing (DB) and datacomm (DC) to the end-user. This bends the curve of 30 years of computer application development to a new direction. The benefits can be many.

However, stand-alone personal computers will be of relatively limited value unless they can be provided with telecommunications capabilities through networks. In this area, the traditional subdivision of the communications facilities into telephony, television, long haul datacomm, and local networking has lost a great part of its significance with the introduction of the new technologies. Wideband capabilities such as satellites, optical fibers, coaxial cables, and the development of intelligent lines through microelectronics alters our concept of designing, implementing and maintaining the communications channels.

The integration of communications services regards not only the high technology itself, but also the nature of the supported facilities themselves—voice, text/data, graphics, live video, mail. A variety of signals, particularly signals coded in a digital form can be transmitted along the same lines and sorted at the user's site.

The so developing spectrum of capabilities promotes both business/industrial and home applications. The implementation ranges from office automation to home service such as Videotex. From electronic mail to telemedicine, the applications have a significant economic and cultural impact as they lead toward an automation of the professional workplace and alter the concept of commuting for communications reasons.

Which are the inventions of greatest potential? The first is the communication satellite. Satellites are now providing telephone and television links to nations throughout the world. In some lesser developed countries the satellite antennae stand next to fields ploughed by oxen. And satellites have the potential of revolutionizing corporate communications both nationally and internationally.

Along with the satellite come tools of the communications process such as low cost earth stations. Planar microwave circuits make it possible to mass produce satellite receiving equipment at very low cost. Such receivers that are inexpensive enough for home purchase have been used in Canada, Japan, and India. Through protocols satellites or other high capacity channels can be shared by geographically dispersed users in a highly flexible manner.

The second most important invention is the laser. While laser-based transmission is still in research laboratories (and has the potential of carrying many millions of simultaneous telephone calls or their equivalent), laser-actuated media have been in operation for years.

In terms of communications, the reference is made to optical fibers, however, we now have laser-actuated memory devices—optical discs. An optical fiber is thin and flexible. It is made of extremely pure glass that can carry a thousand times as much information as a copper wire pair. Optical communication fibers can be packed into one flexible cable. Similarly, the helical wave guide being implemented can carry 250,000 or more simultaneous telephone calls in digital form over long distance.

The third most important development, the millimeter wave radio, permits communicating at frequencies in the band above the microwaves and can open channels much greater than all the other radio bands combined. Chains of closely spaced antennae distribute these millimeter wave signals. Other similar developments include the cellular mobile radio which interconnects telephones or other mobile radio devices in a city and video telephones where users may hear as well as see each other.

The fourth development is cable TV (CATV), which is based on coaxial cables, having a potential signal-carrying capacity of hundreds of megabits per second (MBPS), which can now be used for text, data and voice—hence for signals other than television. This opens new horizons in communications, particularly in terms of broadband local area networks interconnecting mainframes, minis, and thousands of personal computers.

Other major developments include computer-based switching centers, packet switching networks, store and forward capabilities, and voice answerback services. Voice answer-back and the push-button telephone set, makes every such telephone station a potential computer terminal.

3.3   A TRANSITION IN APPLICATIONS IMAGES

The present challenge is to institute a transition in application images that will allow us to make the best use of available technology. Capital goods and consumer products manufacturers have led the way in this evolution. Microprocessors have invaded an impressive array of products, from TV tuners to electric clocks. A 4-bit processor today can be obtained for less than $1. It contains the minimum amount of intelligence, and it is a small fraction of the end-product cost.

As far as the way we view the new technology is concerned, we have gone a considerable way from the expensive controls placed in million-dollar turbines. Both the cost of the intelligent engine and that of the product justifying its usage have dropped to fractions of original reference. Furthermore, some products contain more than one microprocessor.

To take one example, in 1984 there are an estimated 500 different products incorporating microprocessors (and in some cases minicomputers) in the General Electric line. This is up from 80 in 1977. More impressive are General Electric's statistics. In 1977, the total product volume with incorporated microprocessor intelligence stood at about 60,000 units, in just five years (1982) this increased to over 1,600,000 units in other words by a factor 27.

Wide change in terms of product design is now starting in the office environment through the incorporation at the end-user site of microprocessor and/or microcomputer support. With the new approaches to text and data that are now projected, computers will help make a widespread transition. However, this change from the data processing center to a distributed office facility will become effective only after being adapted to the users' specific needs.

For instance, while the keyboard is the classic tool for secretaries and reporters, it is an obstacle in other occupations. The alternatives are the graphic tablet and voice data entry.

Voice data entry lowers both data capture costs and training costs. It may also be the only choice for occupations in which freedom of movement is essential. But is voice data entry ready for practical application? For a structured information environment, the answer is ''Yes!''

Lockheed has used voice entry in a hybrid assembly area since 1979. It has helped to track materials and stages of assembly, and to enter product-assurance test data. A 49-word vocabulary includes digits "0" through "9," the letters of the alphabet, and such commands as "recall" and "kit."

The new generation of voice entry systems will be user-friendly and will be aimed at managers and clerks, not only assembly line workers. Still, under current technology, graphic tablets are more flexible and their templates can be designed by the workstation. The means are more than ready and they can be implemented at the managerial level as well as in day-to-day operations, that can be used for predefined text/data, and as graphics editors and instruments for color shapes and high resolution graphics.

Graphics tablets are low cost ($150 to $750 per unit) and can be attached to the personal computer at the workstation. Even though this is necessary, it is not enough. Workstations should communicate.

The concept of local intelligent devices integrated into the desk and having the ability to communicate with one another in a way unseen by the end-user radically alters the usual way company management looks at computer services. Traditionally, computing services include the service bureau, software firms, systems houses and computer consultants, and integrated systems suppliers, such as companies in computer-aided design.

As far as the end-user is concerned, the largely centralized, mainframe-based computer services of his company greatly resemble the Service Bureau chores. The direction centralized processing is taking can thus be judged by examining the pattern of growth in the just mentioned industrial sectors—traditional bureau services are slowing down markedly while professional services are showing a spectacular upturn at about 27 percent yearly.

While processing companies grew by 17 percent to 18 percent in 1982, the growth of companies producing packaged software products fell between 31 percent and 49 percent. Traditional service bureaus suffered particularly in the batch area. The batch market share is expected to drop from 22 percent in 1980 to 8 percent in 1987. Yet, batch has stayed around so long as the way of doing data processing that it generated a tunnel vision where the only problems that seemed legitimate were those that could be answered through "established," and largely obsolete, methods. For many companies, sticking to the past sets up a reality distortion complex.

Whether we talk of information systems or of day-to-day operations, the average manager needs routine—needs things done the same way. This reduces fear, the number one issue tormenting mankind.

The person who is a cog in a chain often does not like to see things constantly being discussed because he or she is threatened by change.

If we do not succeed in breaking this vicious circle we will not develop in our work. Such a situation eliminates uncertainty and, therefore, risk. But when this road is followed, the result is stagnation and the repetition of old mistakes.

We must look at office automation (OA) as the opportunity to withdraw as many subjective elements as possible from the office, as a result of the unreliability of human operating units, substituting such entities with computer-based solutions. And we must do so starting from the areas carrying the greater weight (an estimated 75 percent of office costs is attributable to the managerial and professional personnel while 12 percent corresponds to secretarial assistance). Emphasis should therefore be placed on managerial producitivity, rather than on clerical a12ks. We have the tools, however, the array of features available on personal computers can be difficult to choose from—spread sheet planning, networking, graphics, database management. The challenge is in enlisting the support of top information professionals to harvest the rewards that flow from improved personal performance.

The components of a system project in an office automation environment will largely include the following:

1. word processing (WP)
2. data processing (DP)
3. private branch exchanges (PBX) to be used as gateways to the public telephone networks
4. voice mail
5. teleconferencing
6. videotex
7. electronic mail
8. intelligent copiers
9. transport robotics
10. energy management
11. air conditioning
12. security control

The basic elements of a study should include WS, LAN, human engineering, and software packages. Correspondingly, the following three levels can be distinguished in a factory environment as the basic parts of a system study:

1. *Administrative*—for product management, manufacturing engineering, production control, quality control, materials, and inventory control are necessary: OA and Computer aided design/Computer aided manufacturing (CAD/CAM)

2. *Production floor*—including production and assembly robotics, workstation-level computer power

3. *Transport and inventories*—this involves transport robots, workstations, and optimization models, among other components

The office system can be customized to become a vital component of the organization. As such, it will need to coexist with other information processing systems, at least through a transitory period; it will support and operate in a multimedia environment; will include prompt and help facilities; provide fast response times; assure transparent access to a community of users and to databases; and support specialized functions.

Such office systems will provide a combination of the best features that exist today in word processing, data processing, and publishing systems. But office automation to be successful requires companies to change their organization and their way of thinking.

3.4 DOCUMENT HANDLING AND THE DATABASE

One of the real problems in implementing office automation is document management. Yet, very few OA systems are able to provide such capabilities beyond a limited perspective.

A document can be anything from a letter, to an instruction manual, to sales and purchase orders, and so on. Standards must be set, corporate and private documents must be separated, and tools must be developed. It is no accident that electronic mail is increasingly oriented toward documents rather than toward simple messages.

Automating document handling is no easy task since paper resists its substitution. However, there are both advanced and short-term solutions. Today, archiving through microforms costs only the 2.5 percent of corresponding paper records. Yet, there exist noted difficulties in substituting the "traditional" means.

The search for methods to handle increasing paper volumes has led to one of the four principal pieces of nonautomated office equipment—the vertical file cabinet (the telephone, typewriter, and photocopier are the other three). We all know that the copier is the most recent invention, but few realize that the next most recent is the vertical filing cabinet.

Just the same, few are aware that filing is the central activity in office chores—even if the concept of filing itself is relatively new. As a result, few people are aware of the role of files in today's office and the developments which have led to it, even if such understanding is fundamental in appreciating the role and impact of databases.

The full benefits of automation and a host of information services demand low cost, error-free databases supporting systems that keep them up to date and synchronized. For these databases we have a number of memory technologies that are presently available or are presently under development. (By 1987, optical discs should be available at 10^{-5} cents per bit of storage.)

Whether we talk of DP, WP or any other implementation, a database is a computer-based system composed of information elements (IE); organized to serve the data, text, and other needs of the enterprise (voice, image, graphs); accessible by authorized entities (people, machines, programs, other databases); and distributed in a physical sense but structured as one logical entity. An

Information Element (IE) is the building block of the database. It is an addressable entity. It can be bit, byte, field, record, file, or subdatabase, but it is usually structured as an object, a page, or a multipage.

Design-wise, the first fundamental characteristics of a database is storage in physical media, to be made in a device-independent manner, based on relational and virtual memory (VM) principles. The access to the IE should be location-independent and based on keywords (direct) or menu. Depending on the application, such access can be implemented with encryption possibilities.

Assisted through a data dictionary (DD), the link to the applications programs should be made in an applications-independent way. The data dictionary must support a unique data definition and identification on a system-wide basis. Database design should also lead to a communications-oriented structure—a message is a file in the database and every file must be designed as a message.

Text and databases should be designed to be distributed. A distributed database requires the following nine features:

1. directory services

2. on-line control capabilities

3. system-wide IE definition and identification

4. synchronization in update and image consistency

5. assured privacy of data subsets (private vs. public)

6. dependable backup

7. on-line recovery capabilities

8. data dictionary services—with distributed data definitions

9. steady administration

Given the distributed nature of data processing/word processing networks, attention should be paid to where to place the control of the database. Experience demonstrates the wisdom of having local control over the database, this control being exercised at an autonomous area level. An area database can only then be good if it is able to map the IE included in the area nodes. Something similar is true of the centralized text and data warehouse.

Furthermore, given that data processing is formatted and word processing is not (as a result DP keeps the files cleaner) it is proper to design for greater text volumes than data volumes. An IBM study brought in perspective a four-to-one ratio between text and data, in future databases. And as a system specialist was to remark: "Word processing is data processing under a fancy name. . . ."

The files in a database can be personal or corporate; local, area, or central. The implementation of microfiles (personal files) and corporate files needs rules: no personal files should be kept at the workstation level; the files should be properly organized and stored in the local database; they should be either shared among users or protected—but managed through the same discipline.

System design should assure that pollution of corporate records is avoided at all times. Journaling and recovery possibilities must be assured in an able manner. There are three other fundamental requirements:

- to get the data only once, but use it many times
- to capture the data where it originates
- to deliver the data where it is being used

All three requirements underline the importance of individual workstation communicating through a LAN. The LAN should have gateways to long-haul networks and remote databases. User access should be projected system-wide and supported by the best facilities technology can offer.

With the database properly designed and implemented, a host of services can be offered. Managers can use the myriaprocessors, its workstation and LAN facilities to send messages to each other. It is more convenient than telephoning.

A staff member may want a functional manager to see a message before going home, but hates to interrupt a meeting. With the personal computer and LAN, he or she sends the recipient a message and when the manager next logs in a list of messages is presented. The messages in the electronic mail facility can then be read immediately or at his or her convenience. Or they can be filed or printed out as desired.

One way for executives to get started with electronic mail is to have their staff deliver information in an organized form, such as by profit and loss information, operations data, and by sales performance information.

Such information can be delivered in easy-to-read form on a workstation so figures can easily be reformatted into lists, charts, graphs, and spreadsheets, in whatever form the executive likes. In time, added features like voice annotation and easy-to-use questioning functions will enable the busy executive to make greater and greater direct use of these tools.

To assist in the speedy conversion to electronic mail within an organization it may help to stop all other methods of fast delivery—such as telex—after a given date. In a certain financial institution, electronic mail was available for six months without spectacular success in terms of adoption by the branch office. But when the director responsible gave the order "from next Monday, telex is no more acceptable, only electronic mail," within a week the service was adopted throughout the branch network.

It is advisable that messages coming from the external environment integrate into the company's electronic mail system. At its Tokyo headquarters, Mitsubishi receives 35,000 telex messages per day. All of them are structured and handled through computers, which makes it possible to process and answer them the same day.

An on-line system involving communicating workstations can also be used for searching the text and database. One of the major benefits is that users are able to turn long documents around much faster. For example, when a contract needs many revisions, annotations and the inclusion of any new text are much simpler.

Calendar services can also be handled through the system. Executives keep their schedules on the machine so that when one has to know where a given manager is, his or her schedule can be checked on the system. In this way, an office is able to run more smoothly, provide faster turnaround on items, and provide a higher quality of work.

Calendar services introduce the user to the concept of timing, critical for understanding a computer-based system, and for facing the requirements of daily tasks. They also assist in coordinating meetings, making the agenda known ahead of time to everyone concerned, and end by improving the results from the meeting time.

3.5 CAPITALIZING ON KNOW-HOW

The design of systems having the capabilities that have just been described will be successful or unsuccessful depending on the know-how invested in them. We need an engineering methodology able to deal with both the dynamics and the mechanics of future systems. A way must also be found to establish good standards and do so early. We must also find better ways to interface electronic systems with the people that use them.

The power of hardware and software to remove the drudgery aspects of work and make jobs more rewarding must be realized. To bring intelligent machines to the workplace we need a better understanding of distributed computer systems. The workstations, we said, must be highly interlinked with communications, and support a wide spectrum of personal and business needs.

The implementation of an electronic meeting capability, for instance, calls for a range of services to be supported at the workstation level such as:

- initiating a session
- identifying the attendees
- having the ability to intervene in the session
- assuring every workstation participating is in attendance
- seeing to it that everybody participating is comfortable with the session
- checking the security measures to be enforced

The implementation of interactive workstations suggests user-friendliness and a reasonable degree of transparency in the mechanics, though user transparency is always relative. The user is aware of the network, but not of the precise way in which the system works.

Restart and recovery is another technical problem requiring solution. The requirements can vary with the level of implementation. At the management level, classical recovery is no absolute need: If a failure occurs, recovery can be done by recalculation. But the security of the database and its protection in terms of access are a must.

In any system study, identifying critical applications can be a delicate problem. Users consider all of their applications critical. A documented analysis must be done and top management should make the final decision on priorities.

The primary criterion should be critical value. How much money will the organization lose if the application goes under? Other considerations are legal, regulatory, and contractual requirements.

Organizations must review backup and retention schedules periodically to maintain efficiency. Legal requirements change, as do the characteristics and importance of the saved data. And it is also important to coordinate the backup efforts of all the processing sites in an organization.

With distributed processing, for example, one location might erroneously assume another has saved shared data. Rules are necessary as backup data and recovery capabilities represent a form of insurance against catastrophies. Protective measures are needed to reduce the chances of disasters. Physical protection of computer centers is a fairly straightforward matter, however, this is not true of the logical security.

Protecting programs and data in a logical sense is a complex task and it is becoming more difficult as terminals and networks proliferate. There is a variety of new devices (both software and hardware) with sophisticated methods that keep pace with the demand for logical security but their implementation requires that a good preparatory job be done on files and procedures.

Within each function, there are a myriad of details. A backup site must have not only the proper equipment, but also the right planning to see itself through.

The right preparatory work must provide for the details of a recovery plan, which include procedural development, identifying resources, setting priorities, and training personnel. Planning for security and recovery should be comprehensive. It is useless to spend two or three years and thousands of dollars trying to measure risks to the penny. The second approach is to do a general analysis and spend the money to do something about the key problems.

On the applications side, the only way to prove a plan is to test it. Testing should be done at the backup site simply to ensure that both the overall recovery plan and the specific programs run.

Any off-site work should include program editing capabilities. No matter how many bugs one encounters at home, there may be twice as many somewhere else. No organization should assume its programs will run at a backup site just because it has the same operating system. Testing will uncover the difficulties.

Yet, not all testing need be large scale. The most comprehensive tests involve shutting down the computer systems and recording how smoothly the processing environment can be brought up at the backup facility.

Security protection is itself a very important issue. It is a fallacy to assume that because the organization has locked the doors people would not come in the windows. Among the recommended security measures are stringent log-in requirements, security analyses of changes in programs and systems, and careful screening of prospective employees.

Furthermore, because even the best defense against fraud and the most complete protection against disaster can fail, we can cut possible losses by having adequate insurance. Several companies offer business interruption and computer-fraud coverage. Some policies can also cover the additional expenses of recovery operations. However, insurance purchases require careful analysis. Premiums can be high, and though settlement checks may be helpful, they are of little solace if the disaster puts the company out of business. The aim of the coverage should be to help the organization get back on its feet.

Not only should backup and security planning be a major concern at the highest levels of the organization, but also the involvement of the legal department, the financial people, the audit committee, internal auditors, and external auditors will be necessary. Such involvement helps the organization become sincere about disaster recovery planning and the security protection issues as well.

Once again, emphasis must be placed on training as a prerequisite to on-line workstations so that the day the system goes on-line, everybody is ready. The one thing no installation wants is unskilled employees putting information incorrectly into the system.

Raising profits and worker productivity while improving customer service are important objectives for any business—but they will not come around on their own accord. Apart the necessary preparation, if increased productivity is the main goal, a performance measurement system is needed. And management must let everyone know where it stands as far as the goals are concerned.

An integral part of this engineering effort is our ability to develop less costly ways to produce software, as the needs for software continue to increase. The key to increasing programmer productivity is a much greater degree of standardization and the development of large and sophisticated computer aids to design and test within those standards.

Along the same line, we need a more effective quality-control technique for software. Software quality is not easy to ascertain in the field, and often even the designer does not really know the quality of the software product prior to field testing. Poor software quality leads to inordinate maintenance expenses. To improve software implementation and maintenance, we must develop a computer-based reference tool for documentation. It has to be as simple as possible, so that anybody can use it. It also has to serve as a reference down the road so that the workstation installations in remote locations can easily look up what they need.

Chapter 4

PERSPECTIVES IN INFORMATION SYSTEMS

4.1 INTRODUCTION

In the following discussion on computers and communications it is necessary to delineate the boundaries that exist between DP (data processing), DC (data-comm), DB (databasing), and EUF (end-user functions).

Data processing involves the use of computers for processing information. The semantic content, or meaning, of input data is transformed, storage and retrieval takes place for processing purposes, and the output data constitutes a programmed response to input data.

Data communications involves transmission and distribution functions where data is unaltered. It also encompasses network control (and on-line maintenance); link establishment, as well as routing, virtual circuit/datagram, flow control, store and forward, error detection, and correction.

Databasing goes well beyond the storage and retrieval of information. First, it addresses itself to a polyvalent environment—data, text, image, voice. Second, it requires organizational qualities such as classifying and identifying the information elements, then using them as the building blocks of the database. Third, it involves management systems approaches to turn database handling into a finely tuned machine.

End-user functions address themselves to the task of making the end-user interaction with the information stored in the machine both easy and efficient. Some of the functions of the workstation are directly intended to answer this requirement. Quite similarly, this is the scope of publicly offered services such as Videotex.

During the last few years, significant developments have taken place in personal computer offerings, both in the second generation of hardware and in the more powerful operating systems running on personal computers. For example, in 1980/81, the less than 64 KBy, 8-bit processor, Apple microcomputer caught the attention of many prospective buyers who thought quite highly of its processing power. Apple II was, by then, four years old. In 1982, the IBM Personal Computer received much acclaim with its enlarged memory capacity and the 16-bit microprocessor. This engine was introduced in mid-1981. In 1983, the Apple and original IBM PC announcements had been overshadowed by products offering bigger, better, faster, and less expensive capabilities—including a born hard disc, 1 to 2 MBy of central memory and 32-bit microprocess. And, in early 1984, Apple announced a Motorola 68000, 32 bit based PC with excellent graphics capability, showing the way in which industry will be moving. During the same time period, videotex has moved from mainframes to minis, and then to PC/LAN. This impacts a major change on the type of offering by reducing costs, increasing the user base, and expanding the applications horizon.

4.2   A TRULY DISTRIBUTED SYSTEM

The developments just outlined suggest that technology has advanced to the point where computer power is economically available at all levels of company operations. Distributed information systems are tools enhancing the productivity of organizations, whether they feature restrained or widespread operations. Intelligent workstations can bring computer and communications facilities not only to every office but also to every desk, and, perhaps, in the future, to every home. However, productivity gains can only be expected if computer-based tools are used without difficulty by authorized individuals, properly implemented for a variety of purposes, and easily adapted as requirements change. Distributed information systems (DIS) increasingly draw attention, yet their meaning is not totally understood. In a way, this is understandable as DIS comes at different levels of detail.

Originally, distributed data processing (DDP) has been an evolution from centralized processing, which in turn evolved from free-standing modes. Then datacomm and databasing facilities have been added.

Though it combines certain strengths of its predecessors, DIS is totally unique in its own right. It represents a break from the past and its implementation rests both on logical and physical premises. The logical functions include the procedural design for channeling the flow of information, and controlling the physical facilities throughout the projected systems configuration. The object of the physical functions is the engineering of the hardware and software devices, to provide a specific level of capability.

The second departure from past practices has been the ability to bring the most capilary level of DIS to the end-user. With one microcomputer per desk, there is a zero degree of remoteness between the computer power and the person it intends to support.

Typically, in the 1950s and in the 1960s information handling was centralized, batch-oriented, costly, and encountered an inordinate number of errors and delays. By the late 1960s there was plenty of evidence that this state of affairs was not satisfactory. Management needed to do something about a steadily worsening situation. It was assisted in opening a new frontier by the system analysts' own realization that new methods were necessary to master the complex information systems of today's offices.

This was the prime reason we have moved away from the mainframe, which is two degrees remote from the workbench, toward the mini, which is one degree remote. The 1970s were characterized by this transition but it was left to the 1980s to implement the zero degree remote system.

In a way, the mini can be likened to the data handling counterpart of small motors with fractional power, which substituted the "big motor" approach with the main drive shaft that we have had in factories at the beginning of the Industrial Revolution. "One motor per tool" found its counterpart in "One mini per office." Years of use and technological breakthrough eventually proved to us that this slogan was not accurate. It should have been "One micro per desk." This is personal computing.

Microprocessor-based, the personal computer has capabilities that exceed those of 1960 mainframes and, in some cases, of big computers brought to the market in the early 1970s. One may even question the purpose of all that power on a chip to serve the workstation. But we find the answer if we consider that with current applications, and in observance of the principle one personal computer for one workstation, one workstation at each desk, this power is necessary.

The personal computer should be a dedicated engine to the manager, the professional, the secretary, and the clerk. The workstation may have one or more microprocessors but more than one person should not share the same personal computer. The object of this investment is individual productivity, both mental and clerical.

Furthermore, the personal computer should communicate within the office through local area networks (LAN) as well as long haul through data communications gateways.

While long haul communications will maintain their vital importance in electronic message systems, significant interest will be channeled to the implementation of local area networks. Among the issues to attract the greatest amount of interest are:

1. the right positioning of a LAN within the context of merging data processing/word processing, realtime operations, office automation, management graphics, and decision support systems

2. the component parts of a LAN: distributed databases, communications faculties, personal computers

3. the results obtained by "avant-garde" organizations that designed and implemented LAN in their daily operations

Personal computers and LAN help make distributed information work for the people who need it. Computers and communications are part and parcel of the tight interrelationships that exist among productivity, job performance, and the need for understanding the services to be derived from computer support—adding computer power to the end user's desk.

4.3 COST EFFECTIVENESS MADE POSSIBLE BY TECHNOLOGY

By bringing in perspective low cost techniques associated with local systems and personal computers, management's attention gets focused on the positive nature of the microprocessor revolution. Both the mechanics and the dynamics of this communications discipline must be under perspective.

Not only do we need definitions for architectural design but also choices on carriers, baseband/broadband solutions and protocols, which are the central component of a man-information communication system. Technical features are necessary but not enough. Management acceptance rests on cost effectiveness.

Cost features bring forward the wisdom of using distributed resources in computers and communications to answer individual requirements. Efficiency reasons further underline this solution. In the last analysis, the critical question for any system is: Does it pay for itself? The answer comes by way of experience. Standalone personal computers are of little value, they just emulate machines that are now dead. Communicating personal computers are truly a distributed information system for the mid- to late-1980s.

Eventually, the same personal computer should integrate voice, text, data, and image/graphics. At the same time, to use this facility to its fullest, a major effort should be made both to train the user population in the new concepts and to conduct missionary work to promote new images.

Clearly, we must teach the end-user how to work with his or her tools. In terms of dynamics, we must open new horizons for him or her. The mechanics make it mandatory to explain to the user what can be done on a day-to-day basis with the personal computer.

Here is an outstanding example. In 1983, one thousand United Technologies executives earning $50,000 a year and up, plus fringe and other benefits (to nearly $100,000 per year), took a three-day course on personal computers. Upon graduation, each was issued a personal computer with video, printer, and other accessories, to use in any manner that seemed suitable. Seventy-five percent of the executives took the personal computer home to continue their education.

In the background of this policy is a major switch in management attitude toward the use of computers as an individual, intelligent workstation. So far, while computers are commonplace at lower corporate levels, they are not routinely used in the executive suite. This attitude must change.

The next generation of multifunctional, communicating workstations will be more useful to corporation executives. They will include not only computing power but also word processing, file access, telephone facilities, and machines that respond to spoken commands. The workstations will be linked through local area networks to other stations, minis, and mainframes so that managers can share data and exchange messages. They will be able to provide a wide range of services including videotex, electronic mail, financial control, strategic analysis, and graphics preparation at the best cost.

Thus, the distributed information system will reflect logical relationships among functional jobs and physical relationships among components. Actual implementations may use a variety of communications network structures.

In the general case, the datacomm work can be effectively done through gateways on a LAN or by the personal computer itself using proper routines. The LAN will communicate with central mainframes acting as large switches and global database engines. Through its gateway, the LAN will also communicate with a number of other established resources, such as minis that were programmed in the 1970s for specific areas of application. The LAN features its own database, while each workstation may have an individual microfile.

The system, therefore, supports not only individual computing but also individual databasing, and is doing so in a cost effective manner. The cost of a workstation is currently at or below $5,000. For an executive, this represents, more or less, a month's salary, and it can increase his or her productivity 10 to 20 percent, recovering investment in less than a year.

4.4   WHAT MAKES A GOOD PRODUCT?

Distributed information systems must be cost effective. If we wish to bring computer power to every desk, providing databasing capability and assuring datacomm faculties, we must also assure that the solution that we implement makes sense, both in terms of technology and economics.

These two subjects are highly interrelated. It is necessary to focus the orderly progress toward the new information systems environment. Success or failure in the years to come will depend on these two concepts.

I do not advocate blindly progressing toward adoption of personal computing. Quite to the contrary, the company, its executives, and its system analysts must watch every step toward a documented, valid, and profitable implementation of the new technology. Typically, such steps in the past had either escaped the attention of the system analyst, because the user was reluctant to identify his or her real problem, or because the analyst did not care to justify computer handling in terms of cost and benefit.

Valid systems solutions start with the identification of first class products and follow through with good projects. As a cognizant executive was to remark: "A good project is when we can do with $1 a job for which we needed $5."

To face this challenge of cost effectiveness, many companies today are undergoing a major transformation; while some are still searching for their new base. In both cases, the prime direction is computers and communications.

The personal computer and LAN strategy that even mainframers are adopting falls under this perspective. Said a senior vice president of an established computer vendor: "We are undergoing a very important change. We were lethargic till yesterday and were caught with our pants down. But we are committed to be in business and to support our customers. So there has been a management upheaval. The whole company is changing."

What bothers the mainframers and minicomputer manufacturers the most is that they have to compete with startup firms that are very competitive. Even those that will disappear affect the perception and expectations of the customer.

This reference is valid from the personal computer to graphics systems, as well as to the upkeep of requirements in terms of workstations, LAN, host communications, and mass storage.

Just because a good product is cost effective, the market drive at this moment is toward the personal computer. Benchmark results confirm that the personal computer is much more efficient and economical for personal computing than the use of timesharing on large mainframes or minicomputers. The following was found in a study reported in *Datamation*:

• Between the large mainframe and the scientific supermini, results in terms of cost effectiveness were about even.

• Tested against the same mainframe, a Motorola 68000 based micro showed 1/16th of the large mainframe's performance (six percent) at no more than 1/70th of the cost.

• An Intel 8086 based engine achieved an even better performance/cost ratio, at 1/250th of the mainframe's cost. It executed the benchmark at three percent of the performance.

If, then, this is what benchmarks say, what would be for the next couple of years a most likely personal computer configuration able to answer the requirements of the workstation? Though it is usually risky to project, the following prognosis can be made:

• Intel 80186 and 80286 microprocessor; or Motorola 68010 and 68020

• up to one MBy central memory

• 30 to 60 MBy hard disc (possibly with cash memory)

• one floppy for data entry, diagnostics and bootstrapping

• eight or more slots, or an alternatively integrated baseband connection (like Macintosh) with the cable substituting the slots

• Unix or MS DOS, Release 5.0 (merging Xenix)

• integrated software, including spreadsheet, WP, graphics, and other functions, particularly personal computer to mainframe links

Supported datacomm protocols will be both bisynchronous (BSC) and TTY. Serial and parallel ports will be native. A database management system (DBMS) will be welcome; and LAN connectivity will be a must.

Typically, this machine will be IBM PC work-alike, with compatibility of packages (on anything not using IBM ROM); will use some IBM PC boards (provided by independent vendors); will be designed for self-maintenance; and will feature at least two types of printers—low cost and letter quality.

In terms of video presentation (softcopy), the character envelope will be better than 10×16 in a high resolution pixel screen. Color, double-size characters, business graphics (thin and wide lines), and smooth scroll windows will be supported. Off-screen controls will involve: bold, blink, underline, and reverse video.

Bit mapping allows mixed text and graphics and other type fonts to be displayed. Each dot (pel, logical pixel) is mapped bit by bit to form vertical or horizontal characters, graphic presentations, and the like. Good video presentation is most important as, in many cases, a full page screen can minimize the need for a local printer.

Characteristic impact media will be a rolling ball (mouse), joystick, graphic tablet, or touch sensing screen, depending on the application. Slowly, with voice input taking hold, most of these will be replaced by voice-actuated devices.

The reader is, however, cautioned on vendor claims, particularly in regard to the "IBM compatible" features. Such claims are false. Because the IBM PC has ROM (which is patented), many programs written for it cannot be used on the so-called "compatibles."

A better term for these machines is IBM work-alikes. There are also *look-alikes* which support little or none of the software written for the IBM PC, but they have many surprises.

A main concern of a small engine as in the personal computer is its pricing. The machine should not only "work-alike," it should also be price competitive.

Competitive thus means *really* being competitive with the prevailing market prices. A company that cannot meet IBM prices: either at the $600 or less home computer class, or in the professional $1,000 to $8,000 range, would do better to get out of the market. In a highly competitive sales environment, one cannot be successful with noncompetitive pricing.

4.5 INSIGHT AND FORESIGHT IN COMPUTER DECISIONS

New perspectives in information systems have been shaped both by technology and by the growing pool of applications know-how. Both are essential for future growth and for demonstrating the need for increasingly using computers and communication tools.

On paper, cost/benefit is rather easy to demonstrate. The price/performance ratio of computer hardware has improved by a factor of 200 each decade since 1953, or better than 15% per year (accounting for every component entering the equation). But computer hardware is only a small part of the picture. We all know how important software is, and programmer productivity appears to be improving at no more than three percent per year.

Unless new software technology can turn the need from programmer training to training users at intelligent microcomputers, the advances in hardware technology will lack the understructure on which to build. Furthermore, we have been aware since the late 1960s that among the factors that contribute to the success of a systems development project the most is the participation of the end-user. This is often overlooked by management because of the highly technical nature of computer systems and because of the attitude that leaves the responsibility of systems development to a computer specialist. Often, the computer specialist has grand notions that he or she knows what's best for the user— forgets that his or her job is to satisfy the user by providing an efficient system.

There is, however, a rational approach to system analysis—valid with computers and with communications networks. We should always keep in perspective that there is an accurate yardstick available against which to match the applications software capabilities and provide estimates of the needed hardware capabilities. The main objectives of computing are foresight, insight, analysis, and design.

Contrary to what has been thought (and practiced for nearly 30 years) neither the automation of numerical calculation, nor the handling of bread and butter accounting data are the main subject of computing. In cases, accounting data can be treated more economically through simpler methods. This principle is just as true as of data processing as it is of databasing and data communications— we need insight and foresight in order to obtain results.

For years, preparing a computer for problem handling meant programming the machine. At last we have begun to realize that computers are not meant to be programmed by the user and his or her specialists—after all, the user doesn't design the computer hardware. Since hardware and software can be procured through commercial firms, the primary focus should be placed on analysis and design. Here is where insight and foresight are paramount. Users have finally begun to understand this issue. As Table 4.1 documents, available software determines the decision of which personal computer to purchase.

TABLE 4.1 Five Primary Factors Determining a User's Personal Computer Choice

Criterion	Percent of Decision
Available Software	26
Company Name and Reputation	25
Cost	20
Maintenance	19
Sales Support	10
	100

The type of software sought is evidently a function of the applications environment for which the PC is intended. For senior management, the following four applications are most important:

- spreadsheet calculations

- graphical presentation (preferably color)

- electronic mail (interoffice)

- communications disciplines including access to databases (the company's own and/or public)

Different applications are required for middle and lower management and for clerical functions. Table 4.2 indicates percentages for the major applications.

TABLE 4.2 Personal Computer Applications

Application	Percent of Personal Computer Usage
Accounts Receivable/Payable/Billing	25
Text Editing	18
Mailing Lists	14
Financial Planning	13
Stock/Investment Analysis	6
Sales Tracking	5
Payroll/Personnel	4
Inventory	4
General Ledger	4
Graphics/Business Records	4
Program Development	3
	100

Personal computer utilization statistics by industry type are seen in Table 4.3, divides the use population of the professional microcomputer into three classes, where it is clearly seen that financial industries lead.

TABLE 4.3 Business Usage of Personal Computers

Type of Industry	Estimated Percent of Personal Computer Usage
Banking, Finance, Brokerage, Insurance, Services	46
Manufacturing, Processing, Distribution	30
Other: Education, Real Estate, Media, Communications Engines	24

Not only is the personal computer part of the cutting edge of technology, but the identification of industries most prone to employ it is significant in the sense that it provides a good account of where investments in software developments will most likely be made in the future. There is nothing better than customer demand to stimulate work.

One of the areas where we will be well advised to apply insight and foresight is self-maintenance. Maintenance, too, has been a stagnant field for 30 long years. Not only did users rely on themselves to program their machines, but they also depended on the computer manufacturer for maintenance.

Computer systems' downtime has been a direct result of this steady dependence on the computer manufacturer for maintenance. Today with the great availability of computer gear, and the fact that the nearer it is to the end user, the greater the reliability should be—maintenance approaches have to change. That's where remote diagnostics and self-maintenance come in.

The new maintenance policy most vendors are now establishing is as follows:

1. On-site maintenance by the manufacturer's own personnel should be done only when absolutely necessary. In this case the vendor will dispatch a service engineer to the site.

2. A remote diagnostics capability is being provided by most vendors as an on-line service. The so created quality database helps both in item No. 1 and in item No. 3.

3. The customer is trained to establish his or her own maintenance program. A parts inventory is at the site and customer-replaceable units are dispatched from one of the field logistics locations of the vendors, as needed.

4. Self-diagnostics enables the user to test the machine components at his or her workstation.

5. If the replaced unit can be repaired, the customer-assisted maintenance program provides for Mail-in. The defective unit is mailed back to the vendor and a new unit is returned to the customer.

In the general case, the customer has the option of purchasing spare units so that he or she can replace them, having only a few minutes of system interruption. A major midwestern American bank that used this method estimates that it will eventually save an impressive one million dollars per year in maintenance fees.

Let's recapitulate this reference. One of the major elements that must change is our approach to maintenance, if we are to be cost effective. This is particularly true in the case of the devices with mechanical components. Such devices take a great deal of abuse.

As computers and communications will tend to be installed in even more demanding environments, a new maintenance concept must be designed:

1. the use of discardable, customer replaceable components

2. more simplified and modular construction

3. components with built in diagnostics that can recognize failure and notify the user

4. fixed time/usage replacement (the component self-destruct at a known point)

Component replacement directly from the vendor should become the exception rather than the rule. It is time to get out of the computer concepts with which we have been living for 30 years, dating back to the punched card era, and into the new opportunities technology offers.

4.6   TELLING THE MACHINE WHAT TO DO

We said that if we do not improve our methods of handling intelligent machines, the costs will be staggering and the efficiency relatively low.

Virtually, all modern programming methodology is based on the assumption that a programming project is fundamentally a problem of implementation: Program design is decided first, based on specifications provided by the user, but it is in the implementation phase that specifications are tested.

To avoid patches, rewriting, delays, and high costs, it is therefore advisable to proceed with a simulation of the implementation after the first release of design specifications has been made. This is written in the understanding that the user may not have a clear and complete picture of his or her needs. By being coinvolved in the simulation the user will thus be able to assure a complete set of specifications, forming a stable base for program development.

Stated in different terms, initial attempts to obtain exact specifications from the user are bound to fail because he or she does not know and cannot anticipate exactly what is required. Often, the user's statement of the problem mixes reality with aspirations rather than provide exact specifications.

Particularly for new applications such as office automation, users have no experience on which to ground their aspirations. Therefore, the way to proceed is by exploring the systems properties through the use of a precise methodology that helps the systems expert to reach concrete results.

During the last 10 years, for instance, structured design approaches have helped assure that the implementation does follow the specification in a controlled fashion, compatible with other projects. While structured solutions are still valuable, they no more offer the range of capabilities modern technology can provide.

Originally developed to support research in artificial intelligence (AI) and interactive graphics, the new programming tools are based on the notion of exploratory programming: an intertwining of system design and implementation.

Artificial intelligence projects have helped the store of knowledge we can presently use, both in terms of methodology and of specific means. An example of the former is prototyping, considered today a valid way to develop applications. Though it may be costly, in the long run, prototyping gives the user a real feeling of involvement, and becomes a time-saver by helping to debug the projected system. In other terms, prototyping both flushes out the critical dimensions of the problem and increases user satisfaction. When we prototype an application, we learn its parameters; we know what resources we have and, given that we have started putting the proper tools in place, chances are that the prototype itself will tell us how long it should take to develop the system. These factors see to it that through prototyping we are really creating a sense of control and ownership on the part of users, and on the system specialists, though for different reasons.

With the new information technology, users are very happy compared to past years. The reason for this is simple—involvement. Involvement and participation have actually helped to turn around the user relationship with the systems specialists and management of the computer operations.

Artificial intelligence is a likely source of programming methodology. Constructing programs is central to artificial intelligence, since almost any intelligent activity is likely to require a set of concepts imbedded in this type of research.

Also, in a way quite similar to basic artificial intelligence projects, the systems programmer has to restructure the program many, many times before it becomes reasonably valid. A variety of programming environments based on the Lisp programming language have evolved to aid in this developmental work.

Exploratory programming was originally developed in a context where change was the critical factor. However, the cost of such programming systems, in terms of computing power required, and run-time inefficiencies confined their use to only a limited range of applications, particularly artificial intelligence research projects.

An example of another broad area that profited from research in artificial intelligence is interactive graphics. The sharp drop in the cost of computers capable of supporting interactive graphics has brought a swift development of applications that make heavy use of interactive graphics in their user interfaces such as menus, windows, and so on.

The real problem is the provision of user friendly facilities in matching the user's needs in particular situations. In a manner quite similar to systems programming, interface design has always required some amount of tuning. The range of possibilities available in a full graphics system makes the design space unmanageable without extensive experimentation.

The *Smalltalk* of Xerox, *Expert-Ease* on the IBM PC, and similar, have been developed to facilitate this experimentation through graphical abstractions and methods of modifying and combining elements into new forms. This methodology is a contrast to conventional programming which restrains the programmer in the interests of orderly development. Exploratory programming systems, to the contrary, amplify the programmer's job in the interests of maximizing his or her effectiveness, providing the programmer with a broad bandwidth in program design and implementation.

Computer-based, exploratory approaches, however, require programming power tools of great capacity, and the ability not to be buried in detail. Able solutions tend to employ metalinguistic characteristics—a technique of program development by writing interpreters for special purpose languages, which is a basic approach in artificial intelligence.

In this sense, the systems programmer develops an application by designing a special language in which the application is relatively easy to state, underlining whatever the designer decides is important. Metalinguistic solutions can be used to describe the following:

- processing elements
- database accesses
- communications capabilities
- user interfaces
- transaction sequences
- text/data transformations

Application development thus becomes a dialectic process, evolving the application language out of the base provided by the development language.

Computer-based tools are valuable in this approach, such being the case of the program analyzer. Its object is to inform that an existing analysis is invalid, so that incorrect answers do not poison the resulting code. A system built on computer-based approaches will also remember the previous state of the programs which it contains, so that changes can be done and undone if necessary.

Computer-based tools create the necessary infrastructure both to help the specialist on the job working on the project and to notify program management or even the end-user. This becomes an important communications tool for flashing out possible failures and incompatibilities and for informing on necessary action. For instance, that a corresponding file needs to be updated.

Computer support lead to interfacing and, because of multiple interfacing, the resulting language may not be most efficient when it achieves functional acceptability. In this case, it is necessary to provide the tools for an optimization. This calls for the availability of performance engineering tools such as a first-class optimizing compiler able to implement program manipulation.

Like the software quality procedure, performance engineering does not work one shot. It is a continuous activity throughout the development phase, as different parts of the system reach design stability and pass the quality control tests.

Chapter 5

PERSONAL COMPUTING FOR SENIOR MANAGEMENT

5.1 INTRODUCTION

A study of managerial performance and of professional activities points toward significant gains in productivity. "Knowledge workers" are spending anywhere from 15 percent to 40 percent of their time in wasteful activities such as clerical tasks, finding and screening information, or expediting trivial issues.

But managerial and professional productivity is an elusive concept. The real question is not how well-justified automated office systems can be in terms of boosting the performance of the decision makers, but whether or not the people filling knowledge-intense posts are willing to work with computer-based systems.

"The executives," said Dr. Bennett of United Technologies, "are not adverse to the use of the computer. In our company, quite a few wanted to go about it—but did not have the confidence." Hence, the pace-setting program United Technologies set up to train senior management on computers and communications was developed.

The seminar takes three days and concentrates on hands-on experience in the use of the computer rather than on lecturing. The introductory topics last about 10 minutes. They are followed by several hours of actual work with the computer.

This program marks a great departure from policies and practices of the past. Until now, most of the major attempts to improve white collar productivity have focused on clerical and secretarial support. The United Technologies effort is the first organized attempt to break through the higher management resistance level.

While in the United States alone an estimated $140 billion is spent in 1985 on purchasing information resources (computers, communications equipment and so on) to aid clerical and other nonprofessional office workers today, less than $20 billion is being spent on similar resources that support managerial and professional productivity. Even worse, clear concepts on how to go about this job have not been developed during the 30 years the computer has been in use.

Statistics indicate that a sizable number of people are employed as clerical workers and secretaries. However, their total compensation is only half of that earned by managers and other professionals. The specific numbers are close to $250 billion for clericals, secretaries and other support people and roughly $500 billion for managers and professionals. With this in mind, it is easy to see that there has been a misplaced focus on support personnel.

Computer-based ways of tracking information automatically are potential substitutes for manual forms and lists. Computers can run our files and that of others. More significantly, at the managerial and the professional levels, emphasis must be placed on the quality of the work to be done.

Cross fertilization is another key reference. Rightly, the group undergoing training in the three-day seminar of United Technologies is polyvalent. Some of the participants are from finance, others from engineering, production or marketing, still others are divisional general managers. There is a product mix not only in terms of grade, but also in respect to formal education, background, and age.

The big change that must occur in the mind of the participants is one of mental images: From polarization on problem solution, to problems conception and formulation. Learning to work with the computer is the catalyst action, not the goal.

5.2 TEACHING THE MAN WITH THE PROBLEM

Computer-supported activities for managers and professionals will be effective only when we teach the person with the problem how to work with the interfaces between him or her and the information needed—typically stored in the database.

Understanding the information, not the form or medium used to store and transmit, is the key to developing a true productivity and quality program. Interfacing takes three forms:

1. Procedural

2. Logical

3. Physical

The procedural is the toughest and calls for new images. This involves rules and policies not only on the types of information that may be collected and recorded, but also, if not primarily, on the use of this information.

What is the systems requirement to transmit information to other parts of the organization? Who is responsible for making and implementing decisions about information needs, for updating, safekeeping, accessing, retention and protection?

The world of recorded information is changing. New procedural solutions must establish what the roles will be. Logical solutions center around the choice of the right software needed to support the managerial activities. United Technologies has chosen MBA, a Visicalc clone with rather extended integrated software capabilities. The rationale rests on the wisdom of teaching the executives one protocol, rather than changing every time a new program is being used in the hands-on experience.

The hardware (an IBM PC) has been wisely chosen to be upgradable. What the participant in the course receives is a CPU with a 16 BPW microprocessor, a video, two floppy disc drives, and a printer. This PC also comes with hard disc, with a larger central memory, and eventually with a much more powerful microprocessor.

Significantly, the company chose the policy of making the PC a present to the course participant. The policy pays off: 70% of the participants take the PC home and work 1 to 1 ½ hours every evening in order to become proficient with the new tool.

These are people whose time is precious. They make over $50,000 per year in cash: with fringe benefits, company auto, stock options and the like, their salaries stand close to $100,000. Even a slight increase in mental productivity will cover, in a few months the cost of the seminar and the PC.

What are these people taught as an introduction to personal computing? From 8:30 a.m. to 17:30 p.m., three days contain roughly 24 working hours. The first part builds on the background material given to the participant prior to coming to the course: objectives of the workshop, the workshop flow, background for the executive's *personal* computer plan, pre-training assessment of opportunities for personal computing, the meaning of the Communication Revolution, and reference materials. This includes such articles as "The Fortune 500 Microcomputers," "How Personal Computers can Backfire," "Electronic Mail Delivers the Executive Message," "How to Conquer Fear of Computers," and "The Incredible Shrinking Microcircuit."

After the Chairman of the Board, Mr. Grey, talks about the wisdom of learning computers and communications, the Executive Personal Computing Workstation is the first subject covered during the seminar, followed by an introduction to IBM Personal Computer: the electronic parts of the PC, the printer, and the modem. Then, how to activate *your* personal computer is discussed.

This leads to a discussion on the special keys: Carriage Return, Caps Lock, Shift, Insert, Delete, Alternate, Control, Home, Escape, a demonstration of Cursor Movement, and practice entering characters. Next comes the very important issue of software: Giving Instructions to the Computer—which leads to an introduction to MBA (the integrated software package). The first step in using MBA is MBA's special function keys.

The senior executive participating in this seminar is expected to learn both the new concepts and the mechanics of personal computing, not the trivial tasks such as data entry. For this purpose, a data diskette is given to every participant along with the data he will need in the exercises. The goal is not to become a keyboard expert; but to spend time on exercises.

The first hands-on experience is accomplished the very first day of the seminar: Using MBA's electronic spreadsheet. This calls for an introduction to modeling and to spreadsheet usage, practice in using the cursor, loading a sample model, moving the cursor using the *Go To* command, examining the size of MBA's electronic spreadsheet, recalculating results, changing data, and entering data.

Emphasis is then placed on the MBA command set, using the blank (B) command, viewing the contents of a cell, using the delete (D) and the insert (I) command, evaluating the structure of formulas, writing formulas, entering a formula, and getting help (when necessary). Following this, the participant is taught how to format a model, determine column widths, format columns and decimals, enter numbers as text, insert a row, enter headings, edit the contents of cells, and enter formulas.

The next subject treated is justification: justifying columns and headings, then, handling error messages, initializing a diskette, storing a model, and printing results. Examples are given through the development (by the participants themselves) of:

- A Salary Planning Model
- A Budget Comparison Model
- A Sales Forecast Model

The evaluation of performance gives the senior executive participating in the seminar a realistic look at expectations from the course. This is followed by an introduction to MBA's Graphics Tools.

The person with a problem is now ready for the graphics exercises. How does MBA create graphs from models? It is explained to him/her how and why performance aids make it even easier, how to display a graph, how to examine the graph description and the title description, and how to evaluate the legend to a graph description. Also, he/she is taught how to print a graph, plot by rows or columns, and work on the principal means for graphic presentation as follows:

- Multiple Bar
- Stacked Bar
- Perspective Bar
- Trend Lines
- Line Graphs
- Area Graphs

The participant is taught the performance aid to help with the multiple bar graph. Also the participant is taught how to select a performance aid to create a graph, and how to keep going with graphics.

The next subject is using MBA's Word Processing Tools. An introduction to word processing is followed by preparing for word processing exercises: the Modeling/Word Processing CRT Display, then the software. Word processing commands include Edit a Document, Print a Document, Combine Cells, Print Multiple Cells, Store Word Processing Documents, Use the Performance Aids. Further, the participant learns how to keep going with Word Processing.

The third day of the seminar is dedicated to the issue of Electronic Communications—including access to databases. The Dow Jones News/Retrieval is covered by way of overview. After an introduction to Electronic Mail, emphasis is placed on The Electronic Communications Model and How to use Electronic Communications to Receive and Send Messages.

As the seminar progresses in a methodological, well-planned manner, the participant is taught how to work independently. Not only is he expected to apply what he has learned right after the seminar is over, but also later on if he does not remember something he must be able to go back to his own notes and find what he had done in the course.

In order to promote this objective, participants are taught how to use *their* new skills. This is done by the means of simulation exercises, followed by the development of *his/her personal* computer plan.

Other references are also important in keeping things running and are included in the third day of the course. These include: How to use the documentation, how to load DOS, how to use DOS with the work being done, idem, to make multiple diskettes; how to copy a diskette; how to protect a diskette; how to use the document log, and how to employ performance aids in individual projects.

No programming is being taught; only the use of the personal computer and of electronic communications. To the senior executive, a program is a package—and the same should be true for everybody else.

5.3 THE TALENT MARKET

The true object of this course is not to teach "another skill" to the person who has made it to the top of the organizational pyramid. It is to open new perspectives to the senior executive bringing the individual who did his university schooling in the 1950s into the world of computers and communications characterizing the 1980s.

This opens the big window toward tomorrow's enterprise and it conditions senior management in a professional life of steady change. The people taking this course have the managerial experience, but the younger professionals just out of college have the modern tools. We must bridge this conceptual difference before it develops into an abyss.

In the course of the last 30 years, new concepts are widening both the mental and the physical environment of mankind, opening up as yet unexplored vistas. These changes are occurring at a breathless pace, and are immensely greater than ever before, affecting the largest number of people ever.

The problems resulting from current changes which are transforming human ecology both in its elements—man, society, environment—and their interrelationships in finance, business, and industry, are threefold:

1. *How to respond to the impact of technology*, managing it effectively, qualifying goals in this changing reality—not only as corporations, but as individuals and society in general.

2. *How to prepare people and let them become accustomed to this tide of change*, participating in this process of transformation and learning to cope with the ever-new situations confronting us all.

3. *How to study, clarify, and establish the new character and aspects of work*, of the workplace, of the work tools, and of the individuals in the decision maker's seat.

Biologically, we have never been guaranteed from birth—that is to say, from hereditary sources—as to what we would become. The results we obtain, the type of personality we acquire, the very nature of the being we call man, are to a great extent inventions of our own.

The person of entrepreneurial qualities is *not* a "natural" creature. To a very large extent, he/she is self-made. We *make ourselves* through education, teaching ourselves ways, means, options, and decisions. In years past, options were conditioned by necessity. Necessity personified destiny: For thousands of years, this has been a mixture of physics, logic, and sociology.

The physical and logical necessity found its fulfillment through manual and intellectual work. Like language, "work" is a typically human characteristic. In the animal, pre-human world, the concept of work was non-existent.

Even among humans an accomplished model of work, as we see it today, did not appear prior until the advent of agriculture. The concept *work* includes a mixture of activities: the utility, the talent, the organization, the social framework, the product, the stability, the repetitions, and more recently the plan and control.

Like all complex processes, work needs *talent*. It also needs a plan.

The critical common denominator upon which the vitality, growth, and profitability of industry depends is human resources. From this fundamental observation we can derive the perspective of a "talent policy":

• To attract and retain the right people at all levels within the organization

• To maintain an environment so that people are motivated to do their best work

• To develop an organizational structure for the most effective allocation and utilization of human resources

This is the most vital internal commitment to the design, production, and marketing of goods. In a post-industrial history, the key to survival is the human factor.

From the farmer to the artisan, the industrial worker, the executive, and the *knowledge worker*, processes have grown multifold. One great aspect of knowledge work is the ability of man to pose timed objectives to be accomplished through tasks of varying complexity.

There are intensified demands for a post-industrial society. Long-range planning is not "long range" because it covers a future time plan, but because of its impact on the organizational capability to look into the future, on the ability to forecast.

The aim of planning is to help us select a feasible course that will take the company where we want it to be—and the drive of senior management to acquire the skills of the new technologies is the right direction. The use of computers and communications as the spearhead of the new technology influences management thinking well beyond the level of simple mechanics. It makes it possible to recognize and define the key elements in the different stages of work process and to integrate these elements into a total, logical sequence.

The able use of computers and communications makes it feasible to monitor the internal and external environment of the firm, in search of problems and opportunities. Fundamental to the effectiveness of this operation is the existence of a master plan and its maintenance in actual, valid conditions.

A master plan should include overall company goals and strategies. One of the strategies should be the ability to accustom management people to the concept of continuous change. Quite often, in the past, 15 years of experience was nothing more than one year of experience repeated 15 times.

There are countless examples of the dramatic results that have been achieved, in terms of increased productivity and reduced costs, through a steady upkeep of management talent. Said Walter Rathenau 65 years ago: "The extent to which the efficiency of an industry can be increased is quite unlimited. Yet, we will not arrive at complete mechanization of the means of production all in one jump. All we can do is to speed up the process to a certain point. For *all mechanization demands reorganization, and this is the result of a stupendous sum of stored-up labor, inventive skill and living capital.*"

In a post-industrial society, management must reach very high degrees of sophistication. Not only must management theory and practice be continually revised and sharpened, but also we must apply the tools technology makes available to create change and set further forward the pace of scientific advance— while through these tools we must plan to retain control of the whole innovative process.

Fulfilling the new enterprise goals requires the recruitment and organization of human resources on a scale which is seldom realized outside our midst, and which is bound to increase. This demands:

•The development of an adequate class of executives and managers able to organize and direct the armies of people working in giant, worldwide concerns, and

•The involvement and participation of these large numbers of heterogeneous people in a common venture—over and above the fact that they may have different cultural patterns, a different outlook of life and progress, and hence different motivations.

We have not yet fully realized how fundamental the human factor is, even in an automated society; and how in a modern enterprise it may become the main asset or the most crucial and baffling problem, according to the place and role it will have in its structure and decision process.

The successful executives of the years to come must be aware that the function of corporations in a modern, knowledge-intense society is rapidly evolving. This function is shifting more and more from that of a traditional heritage, maximizing the output that can be produced by the combination of a given amount of resources to obtain the highest profit, toward a new one in which the profit motive remains, but is conditioned by other imperatives:

- The preservation of key resources
- Mass as a problem solver
- The technostructure

The technostructure uses high technology to amplify its power base and to promote the network of critical points which it has created: mixed decision centers, executive agencies, planning boards, think-tanks, and their operative dependencies. These are the real backbone of a modern organization.

The following are some of the other trends and developments that financial and industrial companies will be faced with:

1. The organizational structure will look increasingly 3-dimensional, through distributed decision centers.

2. The traditional organization pyramid will become more of a participative structure, somewhat like a truncated cone.

3. Management will reorganize to create smaller, more independent, less structured units within the corporate framework, with reporting relationships to a management hub at the center of a wheel rather than the traditional pyramid.

4. Important decisions will be located within a small top-management group, with greater participation throughout the organization in less important decisions.

5. Goal-setting, regular performance appraisals, and career counseling in which subordinates are asked to state their own desires and evaluate their own capabilities, will be implemented to an increasing extent.

6. The future executive will be more concerned with being a part of a team of communicating decision centers with fast access to information resources.

7. The relative stature of all functional management will decline in relation to program management.

8. Far greater emphasis will be given to internal management education and training programs, with computers utilized to build better managers more quickly.

9. Computer- and communications-based management information systems will become far more important and will be substantially improved.

10. With computers and communications providing the connecting link, greater emphasis will be placed on individual freedom, initiative, and achievement.

Thirty years ago, J. Robert Oppenheimer said that one thing that is new in our time is the prevalence of newness, the changing scale of change itself, so that the world alters as we walk in it. But now we are only starting to realize the exacting qualifications of the executives able to face the challenges of our contemporary and future society.

Talented men and women must have the wisdom and capacity to fulfill four fundamental functions all at the same time. They must be:

• *forecasters* who endeavor to devise all possible futures consistent with the present situations, trends, and policies.

• *planners*, insofar as they have to devise alternative, coherent, and feasible paths along which outcomes and goals may be reached

• *decision makers* who do not rely on intuition or flair, but who are insatiable in requesting information and capable of making sense out of it so that their decisions are based on evidence and generally point in the right direction

• *able users of technological tools*, employing the best that science can offer, relying on these tools to keep themselves ahead of competition

A manager's imagination and experience are constantly tested against the difficulty of the real world. The feedback from action may be determinant in the search for and creation of new values: traditional ones being eroded by change, and becoming out of step with the new conditions; values within which to frame the growing uncertainties and complexities of our ages. It is here that the talented individual's stature can be measured.

5.4 A LIFELONG LEARNING EXPERIENCE

Let's now see how this policy is put in practice. We spoke of an executive program. It is not in these three days of training in computers and communications that a senior manager's world will change. But a manager's world will not change unless these three days and other courses are offered to him.

The typical senior manager taking this course is about 50 years of age. (The range being 37 to 65.) Participation is voluntary with support from the top. Each division of United Technologies takes so many slots in the program and schedules its people for them.

In terms of grade, the participants range from department managers to vice presidents and general managers. The program is designed for them and the contents of the program are important. Horizons in information technology were treated in the first offering. The unanimous response has been: "Take them out and teach us how to use the machine."

A company program in computers and communications is no place for academic courses. Its goal is to close the computer-management gap which has existed for 30 years. Now this gap must be overcome. All management levels should be trained, and it is no minor undertaking—even in a large organization such as United Technologies—to see through 1,100 senior people.

To be effective, training must be done in small groups. At UTC training involves 15 participants at a time in a room with an equal number of PCs—keeping one or two machines free in case of failure, so that teaching is uninterrupted.

These courses are given serially: Monday to Wednesday, and Thursday to Saturday, two groups per week. Some 60 to 90 senior executives are trained every month. It is a lifelong learning exercise.

Teaching in the classical way is kept at a minimum. The instructor teaching the course has a PC. A local area network (LAN) connection is used to demonstrate to each participant what the instructor does on the screen. (An initial idea of using a large wall screen was dropped in favor of interactivity.)

What happens after the course is over? United Technologies statistics indicate that an impressive 75% of the participants take the PC home to practice. Once they get more comfortable, they bring it back to the office and use it there.

While the program is offered by the corporate data processing, the office is at the local division. Hence, local DP picks up support at this point. Applications packages and further communications protocols are examples of local support.

Corporate policy aims to enlarge the base of applications. Since a steady stream of programs is currently offered to the market, a corporate committee has been set up to evaluate the packages which come along and to screen junk. Programs with wide general applicability are bought in volume and at a discount.

The IBM PC was selected for the managerial workstation "because it was there. . . ." It is configured with 512 MBy central memory, graphics capability, and printer. (Color was originally planned, but it is not supported by the MBA package.)

The management WS features a high resolution screen for graphics—with flat screen and multiple windowing the solution projected for the future. As stated, the software is chosen for document integration, using virtually the same protocol over the whole range of applications. No unique solution should be selected per function.

As the users get accustomed to the access of publicly available databases (Dow Jones and so on), new applications will be added, and with them new tools. Mailbox is available for electronic mail and the same is true of personal scheduling routines.

Calendar services and electronic mail are among the pillars of the executive workstation usage. Nobody will buy a PC to do strictly one or the other application. But when the computer-based devices are there, both of them become basic applications. Scheduling must include "open spot" evaluation in a manager's own timetable.

Carrier, for example, has 1,500 EMail (Electronic Mail) terminals operating. The president of the division sends evening messages to his immediate assistants and expects them to learn them before going to work next morning.

"Information systems," Dr. Bennett underlined, "are not useful till all people are on them." Correctly, management has seen to it that the program offering has been polished in the most careful manner. The computer specialists were the first to take the course, both to weed out the bad grains and to create a common background.

No experimental tools should be put in the hands of top management. The specialists should take the bugs out first: A specific reference is voice integration with text, data, and graphics. There are some prototypes, but they will not "sell them" to management at this point in time—maybe by 1986.

Is there evidence of a sharp increase in management productivity? "Not directly," the responsible executive answered, "but there is evidence that the managerial workstation helps to eliminate waste:

- trying six or seven times to set up a meeting,
- the telephone tag,
- the inefficient transmission of a variety of messages in a colloquial way "

There is also evidence that *personal computing* takes hold. The majority of the participants puts in an extra hour in the evening at home to master the computer subject.

Corporate DP/MIS sees through the company's 1,100 top people. For lower levels, the divisional DP must do the job helped by the division's managers who participated in the top program.

Throughout both the corporate and the divisional offering the message is clearly given that there are things to do with technology today that are just a shadow of what we can do tomorrow. But in doing things with the tools at our disposition, education is the strong issue—the key.

Putting computers at the top management level today may not make the highest economic sense, but it does make the greatest strategic sense.

Thirty years of experience with computers teaches us that we cannot automate a department from the clerks level up. We must start at the Corporate Office, set up a pilot to figure what sort of a workstation is better for the job, implement a solution, *then* generalize it. Education does not only make the person able to use the tool—it also spreads to gospel.

5.5   TOP MANAGEMENT'S WORKSTATION

The concept behind the design and implementation of workstations for top management is an answer to the problems the senior manager faces today. The focus by bankers, for example, should be not just on strategies for controlling the highly variable spread between borrowing and lending rates, but also on effective planning for the entire balance sheet.

Most critically, the senior executives must determine how their institutions' overall management process should be organized, equipped, and adjusted to achieve ongoing success both in meeting asset/liability management objectives—and in surviving in front of intensifying competition.

The information system the manager has at his disposal, and which he accesses through his personal WS, must help him define and describe asset/liability management, with respect to: rate sensitivity and interest rate risk. This computer-based system must be of assistance in:

- Forecasting and scenario construction
- Asset allocation and credit risk
- Liquidity and funding
- Capital requirements
- Monitoring control techniques

Computers and communications should promote the senior manager's ability in the planning, implementation, and control process, for matching the mix and maturities of assets and liabilities in ways that maximize net interest margin. This is where the spreadsheet calculations, of which we have been talking, are of great help.

The key concept is the coordinated or simultaneous management of both assets and liabilities. This is promoted through computers and communications.

The management process works primarily by controlling the gap between rate-sensitive assets and rate-sensitive liabilities, that is, the differential between those instruments which can mature or be repriced upward or downward within the next 90 or fewer days. What the computer does is to bring in a timely manner the manager's attention to the facts. It doesn't event any facts.

Computer-based models also focus senior management attention on the volume and mix of these instruments. But, while the asset/liability management process addresses chiefly the relationship between rate-sensitive assets and liabilities, other balance sheet considerations are also crucial. Spreadsheet calculations help with that.

The same example helps identify the benefits to be derived from the communications features of the executive WS in providing rapid, analytically-based communications and processing capabilities:

- One of the implications of more exacting asset/liability management is the availability of timely, continuous, and detailed information from both inside and outside the bank.

- This demands better communications systems and connections to databases for a more coordinated decision-making process, money market operations swift rebalancing of the funds composition, and asset mix.

Access to knowledge bases can help tremendously the planning function and development of profit plans. It makes it feasible for the bank to structure the balance sheet in a documented way, and to anticipate the determining factors of the net interest margins: rate sensitivity, asset pricing, and funding costs.

Expert databases enable the senior manager to forecast noncontrollable factors such as core deposit volume and plan noncontrollable factors, especially the investment portfolio size, mix, and purchased funds volume.

Having an effective financial information system is the basic requirement for constructing the profit plan. With the aid of such a system, management can forecast the volumes of various asset and liability categories through analysis of charts of accounts and average daily balances.

With timely and accurate information along with statistical projections, judgmental management inputs and simulations, the individuals in charge of the asset/liability management function can establish the rates of interest expected to be earned on earning assets, and the rates expected to be paid on interest-paying liabilities.

This part of the process may involve several independent estimates of future rates, followed by a consensus opinion on these forecasts for strategy planning purposes.

From that point on, committee functions are determined. The committee members bring into perspective the wisdom of a computers and communications-based operations room for top management which may be able to help in four important ways:

1. Storage and retrieval of text, data, and graphics about a company's own operations and the external market trends/developments. During a committee meeting, the manager can retrieve any information which is in the database within a few seconds. With up-to-date key information readily available, top management can make decisions based on facts.

2. Access to publicly available databases, which contain a wealth of data at a reasonable fee. There are 3,300 public databases today available in the United States. Both their number and their content is steadily growing.

3. Strategic planning and budgeting. Through a communications center, top management can reach directly the different sectors of the business both to establish plans and to review them. An online access to the logical processing capacity of the organization means that managers can examine the consequences of alternative strategies, develop new plans, and revise their budgets during the course of a meeting.

4. Control of operations can be achieved also online giving a dynamic picture of sales, costs and profit trends through charts and graphs. The computer-supported method of presentation makes it easier for top management to analyse results and maintain a close control over operations and also to exercise corrective action.

Both for strategic and for operational reasons, the asset/liability management committee can develop several alternative scenarios through "what if" analysis. These scenarios typically incorporate such variables as expected loan demand, investment opportunities, core deposit growth, regulatory changes, monetary policy adjustment, and the overall state of the economy, in addition to interest rates on particular sources and uses of funds.

There is, evidently, preparatory work to be done: formalizing the parameters of strategic and operational considerations in an asset/liability management policy statement. It is wise to:

- review this policy in relation to the overall management process
- develop procedures
- act as a link between line management and the corporate planning function
- recommend policy changes when conditions warrant

Currently, the primary concern might be to insure sufficient liquidity. A few months down the line the focal point might shift to strategies for acquiring reasonably priced short-term money over a long period of time. Still later, the critical concern might be to develop adequate capital to support earning assets growth expansion into new lines of business.

One of the preparatory steps is to divide loan, investment, and deposit data into fixed and variable-rate categories—and to provide expected loan, deposit, and investment maturity schedules along with appropriate maturing rates and yields.

Such data can be used to generate funds gap reports, which itemize variable-rate assets and variable-rate liabilities, total the item dollars of these two categories, and then subtract them to determine the gap or differential. The same procedures could be followed for fixed and nonrate funds to give a picture of overall average balances for the balance sheet items.

These variable and fixed-rate data can also be formatted into interest rate sensitivity reports as well as mix-spread analyses. Reports formatted in these ways show the effects of anticipated interest rates and of the volume and mix of asset and liability items on net interest margin.

In conclusion, learning personal computing is only part of the job. Just as important is the infrastructure which should be developed. But *unless we give top managers new images and new technological tools—the organization will stay in the old rotten track, and it will decay.*

Part Two
PERSONAL COMPUTING

Chapter 6

PROMOTING OFFICE AUTOMATION

6.1 INTRODUCTION

Microcomputers for use in the office sell so well because they are inexpensive enough to be financed within departmental budgets. The most compelling reason for the success of the PC seems to have been its ability to boost productivity. Even a recession economy can increase the sale of productivity-enhancing products such as PC, LAN, and can promote their integration into an interactive environment of workstations.

A Booz, Allen and Hamilton study indicated that the implementation of current computer technology could save office professionals 15% of their time—thus recovering the cost of the PC-based workstation in one year. Another study pointed out that 55% of the surveyed corporations plan to acquire and implement the latest equipment technology can offer.

The following can be said in judging the total market potential: There are 83 million homes with TV sets, 54 million white collar workers, 26 million professionals, 4 million small businesses—in the U.S. alone. These sectors alone on a one-person PC basis can provide a potential market of 176 million units—double that amount if we look at market potential worldwide.

What networks of railroads, highways, canals, and shipping lines were doing in the last century, computers and communications are doing in today's and tomorrow's economy. But to be profitable the Office Automation (OA) effort should rest on new concepts and new bases. This is not the typical case. In many implementations, the system thinking behind them is quite old. People try to get new experience with old concepts. It is to be expected that such approaches will not work.

It is quite understandable that organizations try to justify the cost of the new gear. Less understandable is the lack of appreciation that a limited scope will not allow this to be so. New perspectives are necessary.

Office automation systems can help us change the way we evaluate capabilities and commitments. In an applications perspective, text processing permits one to solve the point of origin collection problems as has never been done before.

We can enter text and data *only once*, then use this information many times for:

1. Functional processing (of the same "data generation"). Sales order handling, the execution of sales orders, expediting, billing, customer accounting are examples.

2. The creation of successive data generations. These are higher level degrees of projection and evaluation. For example: sales forecasting, inventory forecasting, production planning, and profit planning.

3. New ways of communication between business entities. Implementing office automation sees to it that public (not private) telephone lines are the backbone of the network; common protocols are used at the workstation level; the usage includes company mail, memos, files, documentation, programs. This is the structure of developing worldwide company networks which can handle document distribution quite effectively and make feasible a wide range of further out applications.

When we talk of OA, we should be thinking of new images involving communicating machines—and not of the standalone word processor. Furthermore, text processing is the first application where users accept that they have to apply packages and *not* reinvent the wheel by writing out their own individual programs.

Only the broader systems view can justify the significant investments in know-how, money, and physical resources required by office automation. Only an organization-wide OA solution can contribute to significant productivity improvements—both clerical and managerial.

6.2 THE SYSTEM IS FOR THE END USER

An adequate system for user documentation addresses the problems of easy access, good quality, versatile publication, efficient distribution, and convenient administration. We need procedures which help solve these problems by providing adequate levels of service.

The user will appreciate easy approaches and timely response. Softcopy should be at a premium. Dividing machine-readable documents into windows permits fully online, subject-oriented access to all pages.

An adaptive user interface can extend flexible viewing control by supporting online subject, title, and data catalogs, and also by providing on-demand output of hardcopy and microfiche.

We must definitely avoid rigid human interfaces and cumbersome output options. To do so, we must visualize (from the design stage) the structure of an automated office, considering the functions that might be used in place of traditional manual office procedures.

Only after this is done, can we talk of specific tools. What are the functional system characteristics we are after?

The answer to this question depends on the level of the organization which we aim to automate. Though there are interrelationships between the specific tools and routines employed at the different levels, each of them has its own most wanted features.

For managers and professionals two issues are outstanding:

1. *Calendar and Reminder systems.* Their goal is to handle volatile information that must be kept up-to-date. Online calendars, tickler files, and follow-up files are some of the facilities that can be provided within this frame of reference.

2. *Mail facilities* through electronic mail and videotex—including aids for generating address lists and other tasks of information exchange. Applications that personalize each piece of mail are attractive candidates for this class, but the most important issue is switching capability and gateways to connect to other communications systems. The same is true of transit databasing, such as provided by store and forward.

For secretarial and clerical requirements predominant aspects are:

1. *Electronic typing aids.* For instance, context editors, spelling dictionaries, text formatters, chart makers, and the like.

2. *Filing and Retrieval systems*, with keywords, text/data security, and enhanced functionality. Increased integrity is a "must" when data is stored electronically.

Not every task in the office chores is a good candidate for automation. Critical questions should be asked to determine whether a computer application is a wise choice. For instance: Is the task a simple, repetitive sequence affecting text/data? Is this sequence executed often? Is there a sizeable body of information that must be maintained? Is this body volatile? Is it frequently referred to by a variety of individuals?

Other criteria include whether the function is too time-consuming to do at all without a computer: the load of the communications requirements, new functionality which must be supported, functions that are not possible without the computer, and whether the task can be enhanced by the use of an interactive man-machine communications environment.

Providing adequate documentation for a computer-based system whose users reside at many sites and share files and programs poses basic problems. Are there effective and generalizable approaches to providing an able answer to these problems? This question returns the OA issue back to its fundamentals.

Office automation systems can be ranked by the level of service they provide. *Service is a performance rather than a construction standard.* The aim should be to capture the user's perspective.

Systems at lower service levels offer their end users less sophisticated responses to the primary problems which are facing them in their daily work.

The most primitive service level involves no online access. Output consists of bulk printing and archival storage in advance of demand. Systems at this primitive level locate and supply documentation in traditional ways such as has been done with batch data processing. Standalone electronic typewriters, copiers, and other office gear which have no online communications capability fall under this class.

A slightly more sophisticated office system adds the ability to find needed documents by online search. But it takes a still higher level of system solution for users to handle their functions on demand instead of drawing from an aging stockpile. Only at that level of flexibility we can start talking of OA approaches—and that is only a beginning.

A higher up functionality, easy user interfaces, fourth generation languages plus the ability to actually consult online to find answers, are examples of office automation functionality. The higher a system's level of service, the more the intelligence which it uses, and the greater the need for interactive tutorial, prompt and help functions.

User documentation and assistance, based on computers and communications, can have many functional answers. At any given level of service there may be different systems whose adequacy of performance varies.

- A level of service, which tells what a system does, should not be confused with quality of service.

- Quality of service reflects how well the system does what it does.

However, at any and every level of sophistication, a user oriented system must be characterized by simplicity, friendliness, flexibility, and control. Simplicity reflects on the extent to which users find the system natural and without unnecessary intricacies.

A system is user friendly to the extent that its behavior is self-explanatory; it helps in tutorials, complements standards features through help functions, and is forgiving user mistakes. The latter is written in the sense that it allows the end user to recover from his own mistakes, roll-back the operation, and repeat it.

Flexibility has two meanings. One meaning overlaps with what has just been said, as to the extent to which the system tolerates errors and lets users adjust the way they supply input or generate output to suit varying needs. The other goes beyond this reference in terms of system structure—and into the ability of modular approaches involving hardware, software, and functionality.

Flexibility should also characterize control perspectives, particularly end user control: the extent to which system actions are started and controlled by the user, rather than forced by system constraints. These criteria should be widely recognized in a rational OA design.

6.3 SYSTEMS STUDIES

Successful approaches to office automation must benefit from specific goals properly integrated among themselves through a systems study. The latter should include: telephones, stored program PBX, communicating word processors, electronic mail facilities, facsimile exchange, communicating copiers, controllers, multiplexers, and above all the implementation of PC-based individual workstations linked among themselves and to other resources through local area networks.

Rational system studies of this nature are not yet the rule in industry, one of the reasons being that current sales of OA equipment are predominantly for replacement purposes:

- An estimated 75% to 80% of the systems installed replace some form of existing standalone text editing or word processing unit with no communications capabilities.

- Another 10% replace typewriters.

- The balance is going into areas where nothing had been done before.

Such statistics have in the background the functions of the person buying OA equipment on behalf of his organization. Less than 5% of the decisions are made by a systems architect responsible for office information systems; 15% to 20% by small task forces, the I.S. manager or a DP/WP coordinator; the 75% or more is made by the administrative director, the office administrator, or word processing supervisor.

Such statistics may, however, be changing. An increasing number of companies look for the capability to integrate into OA an impressive array of functions which spread from databasing and datacomm to telephone service and microforms. Recent rational studies attack four types of problems, with particular emphasis in the documentation domain:

1. The *access problem* involves finding answers in available documentation to a user's requests for information. Able solutions are fundamental to any office system. They are also a prerequisite for an efficient approach to the following steps.

2. The *transport problem* concerns transferring copies of documents to those who need them. At computer-based solutions documentation should reside in machine readable form. Because access and transport solutions are interleaved, the extent to which we can exploit this link will be instrumental in shaping up an effective computers and communications approach.

3. The *quality problem* calls for providing users with readable, well-organized, up-to-date text, data, and images. Any OA system should be able to assure

clear and current documents, preferably in softcopy form. Schemes with rigid format constraints or long delays pose built-in barriers to document quality. The OA system should promote clear content and rapid revision.

4. The *administration problem*. This often results from the means adopted to solve the primary technical issues, but its impact goes beyond the scope of document delivery. The administration problem requires the achievement of a workable, economical framework to support the choice of access, transport, and quality. No matter how easy the access or how versatile the text/data transfer, an adequate system must also be economical, efficient, and maintainable.

Other administrative problems regard prerequisites established by management. For instance, security/encryption capabilities.

- Often a database is penetrated with the help of insiders who provide enough details to make unauthorized access possible.

- Most corporations transmit large volumes of data using wireless technology—but few understand how vulnerable their transmissions are to illicit monitoring.

Usually computers do not contain devices to safeguard sensitive data. In addition, many organizations ignore the simple protective steps against text/data thievery.

Yet, encryption has been used to protect messages as far back as ancient times. The Spartans developed a simple ciphering device called the "skytal," a cylinder wrapped with parchment tape. A secret message was written across its full length, the tape was then removed and worn as a belt by a messenger. The recipient could read the message by rewinding it on a cylinder of the same diameter.

At the present time, mathematics and electronics have made possible the development of rather reliable encryption devices. Cryptography alters data to make it useless to adversaries or unauthorized listeners, thus preventing the possibility of injecting false data into a communications channel, altering a message in transit, and so on.

Modern cryptographic systems transform information using specific algorithms controlled by a unique number or bit pattern called an encryption key, thus producing enciphering sequences. The recipient can decipher only by using the appropriate key which (it is hoped) is not available to intruders.

The question of text/data integrity is vital to most users, but few organizations have attacked this subject in an able manner. Valid solutions necessarily include: engineering support, the testing of alternatives, diagnostic facilities, and improvements on overall operation to safeguard valuable data—which is after all a corporate resource.

Management should contractually imply the vendor's responsibility to assure the observance of promises given at the time of the contract. While this is generally true of turn-key jobs, it should cover the whole range of computers and communications systems.

It is not enough to say, or even decide, that paper as a record keeping medium should be eliminated. It is also necessary to study how the electronic files will be maintained and administered in the most efficient manner. And it is just as important to understand what is necessary in terms of organizational change, and what is involved by way of preparation. Administrative recommendations should range from dependable line controls and efficient switching procedures to interactive capabilities, online oriented activities, workstation implementation, reliability insurance, and the sharing of technology's advances with the end user (more facilities, lower cost).

Administrative procedures should spell-out in advance of system implementation the organization's sensitivity to diagnostics and service ability. Today, the aim to locate hardware, software, or communications faults takes too long and offers too little in evidence, but the trend now is toward remote diagnostics and self-maintenance. Both are necessary as hardware repair time is usually a fraction of the time it takes to locate the fault.

The administrative side must steadily upkeep the company office system requirements, detailing by means of indicators what is meant by the broader, management established goals, such as:

- increased professional productivity
- flexibility and growth in stages
- fitness within the existing organizations
- ability to tie into existing data processing applications

Administrative procedures must see to it that monitoring aids, both for software and for hardware, are improved. Furthermore, since change in procedures, systems, software and hardware have long lead times, the systems study should assure that the new structures are able to coexist alongside the old.

It is part of the overall administrative duties to look after the orderly introduction of computers and communications into every field of company activities, abandoning the often evident level of relative resistance to accommodating change; to profit from the advancement of technology; to use decision support capabilities; to convert to exception reporting, and to offer new, profit-making, computer-based services.

6.4 EMPHASIS ON COMMUNICATIONS

An office automation system relies heavily on three resources: the microprocessors integrated into the individual workstation; the associated microfile capability; and communications. Many communications paths exist within the interactive computing system.

- The secretary communicates with the boss.
- The manager or professional with his/her peer.
- A database with another database.

We can look up files owned by someone else (if we know the password), obtain copies of those files, and send copies of files to individuals through the computer network that links their locations. When communication involves the use of an interactive computer system, the amount of paper movement in an enterprise is reduced.

Internally generated correspondence accounts for some 90% of the paper that crosses our desks.

Electronic mail and sharing of files cut significantly the amount of generated paper. They also speed the exchange of information.

The emphasis on communications does not diminish the role of personal computing and databasing. It enhances it. To communicate effectively, we need editors and other facilities for:

- locating and changing text or data on the basis of context
- making global changes
- moving and duplicating blocks of information elements

Effective communications also imply the ability to provide added functions, change or specify file characteristics, and enter codes for graphic sets and characters. Facilities needed for editing text documents are available in most present-day editors largely because of the communication requirements observed by the developers.

An attempt to define what is good full-screen support, suggests ways by which the user can directly alter the contents of any line he sees on the display screen. This impacts on the structure of a linkage to the information elements in the database. And since a message is a file stored in the database—and any file should be structured as a message—the process we are describing carries all the way to the communications side.

The inverse is also true. Solutions given to the communications problems impact on the database, and by extension they effect end user presentation both in terms of input and of output. While text processing facilities may be looked at as a specific example of personal computing, and text/data transfer as an example of communications, the two are highly related:

- Text processing facilities include software and hardware able to provide assistance in administrative/secretarial jobs.
- Communications facilities allow the dissemination of timely information to all workstations on the system.

Processing facilities will be used to prepare a document for communications purposes. At the same time, the availability of good online communications promotes processing at specialized workstations. There is a whole range of common computer-based services to WP, DP, DC. Help aids, query facilities, and online databases ensure documentation is current and immediately accessible.

These references are part and parcel of the specialized tools which exist to enhance individual productivity. Calculation aids, quick lookup commands, electronic mail, voice mail facilities, and menu-based structures are other examples.

The underlying reference is that we need tools to support document processing typical of most organizations. Input/output functions are important features to consider in any database and datacomm evaluation. They should allow the user to establish specific parameters regarding text/data input, to input edits, provide formatting of data, or even create more complex forms, such as an invoice or statement—at user request.

Inquiry functions are an important aspect to the system's ability to function as an online information source. Updating capability can be a good measure of flexibility.

Both for datacomm and databasing purposes, a considerable amount of study must be devoted to understanding the paperwork process within the company. Conventional business correspondence, such as letters and memos, should be sampled to determine where they were coming from and where they were going.

The pattern that emerged in one study, on both the incoming and outgoing side, was that a substantial amount of paper stayed within the company: 75% of the incoming letters and memos originated within the company and 82% of the outgoing documents remained within the company.

Communications requirements are often burdened with broadcasting to a distribution list. At the same time, and for the same reason, this is an opportunity to relieve the data load of the traditional office chores—through online communications.

An office systems study, for instance, found that a lot of time was spent in copying: For each original, six copies were made on average. Most of these were machine copies and the time devoted to making them was highly unproductive time in that it involved traveling to and from the copiers and waiting for them to become available.

As it cannot be too often repeated, the DC component of a system is an integral part of the system's structure. As such, it influences the other components and is as well influenced by them. Placing emphasis on communications does not mean only to look after the transport activity. It means much more than that.

Even this more limited scope will involve functions such as: logging of mail, control of the status of work in process, disposition, security for control over document access, modification(s), filing, duplication, destruction, audit trials of access, and so on.

The point to be properly conveyed is that the communications study should properly examine issues related to *document capture*. This is particularly valid for material originating outside of the system which must be converted into system documents. Incoming mail, magazine articles, photographs, handwritten notes, charts and graphs should be entered into the database and the communications channels using scanning devices.

A similar reference can be made in regard to *document creation*. Entry, edit, correction of text, and so on, may be required for the preparation of correspondence, reports, and forms. A system facility for the definition of such forms is needed—to be followed by the study of transport and delivery through modern communications means.

Videotex has been one of the aftermaths of the Knowledge Society, and now promises to become its agent. The innovations which it affords start finding their proper place in industrial, business, and financial life. Electronic banking, point-of-sale transactions, hotel and travel reservations, computer conferences, facsimile distribution, image processing, interactive graphics, text editing, document distribution, electronic information services via television, are but a few examples.

Other technologies will be needed to support such parallel activities as voice message storage and redistribution, economic storage media of mass capacity, security locks and a variety of interfaces, the substitution of the software the way we have known it so far. And while some of these developments have become available, others are still in the laboratory.

But the use of technologies such as videotex poses prerequisites, the most important being adequate preparation. Access to the viewdatabase, routing frames and infopages, the page mechanism, format, content, editing, text and data entry, file handling procedures, and accounting for system limitations are some of the actual requirements the user must be aware of.

This places emphasis on other closely related DB/DC aspects such as file, search, and retrieval. Electronic filing of documents during active use should include capabilities for creating multiple personal files, establishing descriptor indexing schemes, and support fully automatic indexing.

An OA system may, for instance, be instructed to retain only one copy of a completed document. Documents could be moved automatically to lower-cost archival storage when current need was ended. Provision for identification of desired documents could be done via indexes, or keywords, alone or in combination, through search queries. The document(s) could then be accessed and reviewed.

The aim in electronic communications is handling transactions and messages. This also brings up the aspects of user education, social convention, cost justification, and network integration. Bringing voice under the same envelope of the message service is a goal which does not need to lie too far—as hardware and software are becoming friendlier to the user.

Companies hesitant to undergo the radical rethink of their internal structures that is required for the effective introduction of the technology, will find the greatest roadblocks in their way. Now, more than ever, those organizations are able to survive that:

- have clearly defined business goals
- are action oriented
- are simply organized
- have lean staffs
- view people as the key to productivity
- utilize operational autonomy
- emphasize the company's strengths and correct the weaknesses
- have a few tight, clearly defined controls
- allow considerable flexibility in other areas

Whether by trial and error, intuition, or design, successful companies have found a particular set of values that make them the leaders in their field, give importance to detail, and through proper planning, are prepared for every eventuality. "Chance" favors the prepared, and preparation means training the human potential. Automation or no automation, the key jobs are done by able people.

The advancing technology accelerated the obsolescence of our knowhow and made lifelong learning even more mandatory than before. Banking used to be a slow-moving profession in terms of innovation. Yet today, the banker who does not train himself in the new technologies loses half the knowledge he needs for his job in the short span of five years. That time lead is down to three years for systems specialists. Technology is a demanding partner.

6.5 ARTIFICIAL INTELLIGENCE TODAY

The field of artificial intelligence (AI) is unusual in the sense that many of its goals reach into areas quite apart from one another. Learning, vision, motor control, obstacle avoidance—are problems which have both been solved some years ago and are perpetually in the list of issues requiring further solution.

For many years, AI was considered something of a laboratory curiosity. Quite recently, however, research has broadened considerably leading to the formulation of models of behavior in fields such as weapons systems, medicine, chemistry, geology, genetics, and many branches of industry.

While intensive funding in artificial intelligence may make possible an astonishing new breed of weapons and military hardware, it will also have significant fallouts in business and industrial developments. Among weapons systems drone aircraft, unmanned submarines, and land vehicles are examples that combine artificial intelligence and high-powered computing. The "automated battlefield" will lean heavily on AI—but weapons systems research also focuses on creating artificial-intelligence machines that can be used as battlefield advisers and command engines able to coordinate other complex weapons.

The principles which we have been treating, are important in two large areas of civilian applications: expert systems, and robotic vision.

Expert systems can emulate human expertise in a well-defined area. General Electric is building a software program that will provide expert advice on repairing locomotives. But expert systems can go further than that, making creative use of errors.

Scientists doing research on artificial intelligence are striving for an ultimate goal: a computer-based analog of the human brain. In the forefront of such effort is a new attention to mathematical sciences.

Applied mathematics are made transparent to the end user by means of computer power. Augmented through machine processing, the use of mathematical techniques yields dramatic savings by reducing the time and money otherwise needed for real world experiments.

Out of such research emerge intelligent decision support systems such as one created for internal medicine. In developing a theory of expert problem solving we usually begin by acquiring a small number of facts. By studying them, hints may emerge as to the expert's structuring of knowledge. Such research may provide clues as to the process by which this knowledge is accessed and used in the course of reasoning and problem solving.

The challenge is to fashion from these empirical findings a hypothesis which can be tested. Based on the study of facts and on the expert's testimony, we can construct a working model. Originally, this is a model of the analyst's concept, which may not bear much resemblance to that of the expert. Nonetheless, the model can serve to guide and sharpen the further search for more subtle aspects on the way the expert's information process works.

Expert systems turn the computer into a consultant by copying the decision-making process used by human specialists.

• Schlumberger has a program that interprets geologic data in the same way a human geologist would.

• Digital Equipment uses an expert program to select the components to build a computer system designed to meet a customer's specifications.

All this is an evolutionary process, leading to a progressive deepening of theories of intelligent behavior. This reference explains why AI scientists consider the availability of powerful hardware and software tools to be essential for the conduct of their research. Typically, AI programs are fashioned in such a way as to facilitate experimentation, with little regard (at the beginning) for the efficiency of program execution.

From time to time, a system developed in the course of this type of research will be found to have sufficiently valid behavior to warrant its refinement and development into a decision support tool. The expectation that commercially viable AI systems will soon become commonplace, rests on this simple fact.

However, AI programs tend to be very large and often require access to very large knowledge bases. Programs that do not call for a large amount of historical knowledge, pose considerable demands for dynamic memory. This is so because they involve searches of large, potentially infinite combinations, characterized by an inescapable explosion.

Artificial intelligence problems are compounded in the design of real-world decision support systems that require access to large databases and search methods permitting exploration of underconstrained algorithms, using a euristic type of reasoning strategy. At many research centers today, methods are being developed for combining the facilities of relational databases and those functional programming languages. This may well prove of critical importance in the development of models able to handle realworld knowledge.

The simpler models will typically include data for forecasting a range of issues of interest to management, from the effect of allocation decisions to how different materials will affect manufacturing costs. As times goes by, managers find themselves confronted more and more with "what-if" kinds of questions and they need tools to handle them.

Computer vision is a different field of AI implementation. Stereo vision, for example, supports the ability to calculate depth by measuring the differences between two images of the same scene as viewed from slightly different angles.

For robots, stereo vision is invaluable for supplying information about the shape and texture of objects in their environment. The key to this process is finding out which elements are the same in both images.

Success follows as much from knowing what to ignore as from gathering finer levels of detail. Strength and ingenuity lie in the way the machine smooths away distracting levels of fine detail in each image by calculating the average brightness of each pixel of light on the display screen (as a function of the brightness of its neighbors). This averaging makes it easier to match objects in the two images. Another benefit is that it makes the device especially effective with textured surfaces where conventional image processing techniques may not work.

Stereo vision equipment marks the places where brightness changes are unusually great. This is displayed as a pattern of contours. Then contours from the two images are matched with each other. The machine measures the differences in rotation and position and calculates information about the third dimension from the differences.

Yet, the currently available AI products are only the first step toward making computers more useful. A host of research is under way: IBM, Digital Equipment, and American Telephone & Telegraph—among others—are doing considerable AI research.

In mid-1983, International Business Machines gave a resounding commercial endorsement to artificial intelligence when it started selling rudimentary AI software for its large mainframe computers. At the same time, a dozen small companies as Computer Thought, Machine Intelligence, and Teknowledge have been formed to do AI work. Many of these are expected to develop more advanced commercial AI products.

Such references help document the fact that artificial intelligence is emerging as an important new commercial market, rather than just as a focus of research and curiosity on the part of universities and the military. Hardware and software sales in this field are already in the tens of millions and, before the end of the current decade, into the billions.

Important applications are developing in office and factory automation as well as in medicine. This is an opportunity area more embryonic, but even more promising than the personal computer.

Expert systems are the first practical application of artificial intelligence— like the telegraph in 1835 was the first practical application of electricity. Since the building of the first expert system 15 years ago, several top projects have demonstrated that man-made systems are capable of expert performance in diagnostic, interpretive, planning, and control tasks.

The object of these pages is *awareness*. While we may not be ready for a certain technology, or a given technology we might wish to apply is not ready for us, we should always be keen in knowing what is coming.

Much more than professional interest is involved in this statement. The systems which we design today must have a life cycle of ten to fifteen years—no less than that.

- This means, they must be able to live to the end of this century.
- To build such systems we surely need to know *what's next*.

If the ten to fifteen years' timeframe seems too long, let's not forget that we still live with software projected in the 1960s and converted a couple of times from machine to machine. *If some information systems today are deadly obsolete, awkward, and unprofessional it is because foresight was not used in their making. Let's not repeat the same error.*

Chapter 7

INFRASTRUCTURE FOR EXPERT SYSTEMS

7.1 INTRODUCTION

The executive office of the future will be an environment in which resources such as computer software, hardware, and communications means are interconnected and integrated at all levels throughout the organization. Decision support systems will involve computer-run mathematical models, simulators, evaluators which are transparent to the end user. Workstations will be of two varieties:

- deskbottom PC interconnected through LAN and involving color and graphics

- video wallpaper for conferencing

Terminals will be used in board rooms and other committee settings to recall and project on wide-screen information from databases.

Senior management must be able to use electronics support to retain effective control over the executive functions up the organization ladder. This effort will be successful if and when the right infrastructure is created, used, and updated.

The infrastructure can be built through basic studies. Exxon spent a rumored $200,000 to figure out when, at what currency, in which country, and what time period they should pay their bills. Also they had to figure out where to buy crude, store it, etc, given current markets. This cost was recovered in two months because of market fluctuations on which they were ready to capitalize.

The emphasis on decision support systems is in realization of the fact that although blue collar workers and clericals have been helped somewhat by automation, managers have not kept pace. Yet, since professional level wages on average are two to three times clerical wages, even a modest improvement in productivity at this higher level could result in substantial economies.

While expert systems are well launched in their way of development, we are perfecting delivery means for management information through such approaches as videotex. This effort involves organizational prerequisites, transport facilities, warehousing for data/text, and user friendly reporting characteristics.

Videotex operates in *real enough time* using existing structures: telephone, and television stations, and networks. It brings text, image, and data to the fingertips of the end user—acting as a catalyst in rethinking organization, structure, text/data collection, storage, and reporting.

Expert systems can use the videotex capability as a carrier. Their implementation at the computing level implies, however, its own perspective.

7.2   A SYSTEM TO ENLARGE THE MIND'S CAPABILITIES

Information which supports decision making for longer range policy and strategic decisions usually rests on key variables which, to a significant extent, can be quantified. Typically, such variables relate to: market sectors, financial goals, and the study of the available alternatives to reach objectives.

This is written in full realization of the fact that most senior managers rarely spend more than an hour at a time on any one activity. They deal with a typical problem in time slots distributed over a period of weeks and months. In a typical day, they cover many different tasks.

Yet, in spite of this discontinuity in allocating time, studies done for reasons of *expert systems* demonstrate that computer based aggregates are useful to the manager. Even if his decision making is often ad hoc and addressed to unexpected problems, it is possible to develop reporting systems which have scope, flexibility, and relevance—if we keep away from stereotypes.

First of all, we must put clearly in mind that decision making involves exceptions and qualitative issues. While plenty of computer power has been available for three decades, this fact wasn't reflected in its usage. Times are, however, changing.

With artificial intelligence further out on the horizon, we have at this time enough expertise to assure that an interactive information system can be developed for senior management and the planning officers, designed to allow:

- Visualization of figures and graphs

- Presentation of alternatives with calculation to answer "what if" questions

- Simulation of the effects of given policies on the results of financial and other types of allocation

- Optimization studies based on criteria established by management

A color graphic terminal would make it possible to obtain very clear pictures of trends; a database gives a simple means of retaining easily accessible archives of the most important results; a LAN helps interconnect executive and secretarial workstations; a color TV projector can serve group meetings, from study groups to management committees.

End user functions can be presented in a user-friendly manner, linking the man behind the machine to a database, supporting interactive capabilities. Interactivity with the elements in the database is important as decision analysis is not static but dynamic—its key ingredient being that of *assessing probabilities* through:

- The formulation of hypotheses

- The concretization of paths

- The verification of options

Verification should not only point out the consequences of a given process in specific situations, but also pinpoint the way of presenting the basic properties of the assessment being made.

There is a very significant difference between text and data processing for operational and for managerial reasons. Table 7.1 demonstrates "why," along four frames of reference: type of data, applications, volume, and response time. Management data should be presented in graphic form. Whenever tabular information is necessary, three orders of magnitude will suffice.

TABLE 7.1 The difference in operational and managerial requirements

Use	Data	Applications	Volume	Response Time
Operational	1. Detailed 2. Volatile 3. Requiring frequent update 4. Current 5. Subject to recovery in case of failure	1. Repetitive 2. Predefined 3. Massive 4. Originally paper-oriented with visualization as a substitute	High	Critical in terms of RT update—but a few seconds delay admissible in reporting mainly positional
Managerial	1. Synthetic 2. Stable 3. Non RT update 4. Historical tendencies, correlations 5. Elaborated data with ample backup at operational level	1. Non-repetitive 2. Not predefined—hence modular, flexible 3. Exception reporting 4. Basically visualization; preferably graphs and color	Low	Non-critical in RT update Critical in • Immediate response • Security/protection • Graphic presentation • Algorithmic support • Directional sense

The infrastructure to support an expert system should account for the fact that management decisions:

1. Are made in the face of uncertainty

2. Take place with insufficient information

3. Deal with the allocation of finite resources

4. Aim to preserve resources or, usually, increase them

5. Reflect the texture of the information environment in which they are made

We can develop *aids* to management decision through graphics and algorithmic expressions. We can assure compatibility between communications solutions, operating processes, databases, and the relationships among the information elements they contain.

The knowhow is available to plan for integrity, making feasible error reduction, assuring journaling, providing for security/protection, and supporting media for instantaneous response. Managers need *directional*, rather than positional data.

Modern systems thinking calls for planning for an integration of goals between decision makers and information providers. Design aims are now capable of responding to the information needs of the users—at each level in the organization.

One of the overriding needs in system design is to assure an easy transition to new systems. We do not wish to upset current structures—until they are replaced.

7.3 PROBLEMS AND SOLUTIONS

Years of experience help document that a recommended development of a computerized, interactive system would: have a time series-oriented data storage facility; accommodate external as well as internal data sources coming from manual and/or automated systems; and provide for narrative and statistical sets of data able to reside in the database under a catalog of available information.

Such systems will be designed to assure flexibility of analysis including spontaneous as well as scheduled types. It will be open to graphics, preferably color, and avail an output facility capable of individual or collective output displays to be run in an interactive manner.

Typically, a system designed along this path will enable the user to communicate directly and quickly with data and evaluate alternatives, making full use of the system's analytical aids. It will also provide features which allow a decision maker to pose and repose questions considered crucial for decisions.

Today, management information systems incorporate facilities making them easy to use, and they are adaptable to a changeable environment through an accommodating system architecture. But requirements go further than that anticipating an evolutionary development cycle.

A modular approach suggests itself. User modules consist of a series of related decision options, and decision processes. The need for information for decision grows more rapidly than most data systems can effectively provide. Modules can act like a prototype of the decision process. They allow quick response and growth in concert with the needs of management.

Modules access a common database and display information personalized to the decision maker's style, patterned to the environment to be managed. Menus also comprise an important aspect. They develop possible sets of choices in order to visualize the alternatives to be analyzed for the decision.

Based on a utilitarian design concept, menus enable the decision maker to select from a table of earlier choices or to develop new approaches for examining the relevant factors of the decision. Decisions concern a growing variety of problems.

Managers face problems, and problems are, by definition, matters involving uncertainty and requiring solution. The expert system, like the expert himself, needs a line of conduct. This is the process one proposes to follow in decision making.

Indeed, under present day technology, the best way to establish the exact formulation of the structure of a decision system is to interview the executive with acknowledged skill in his line of activity and find out how he thinks, acts, and reacts.

In many cases, when confronted with complex situations, a manager's way of thinking is euristic rather than algorithmic. Heuristic is an approach to the discovery of a problem solution, through the making of plausible and fallible guesses as to the best thing to do next. By contrast, algorithmic is a completely specified solution procedure which can be guaranteed to give an answer if we follow through the prescribed steps.

If an expert system is supposed to follow the track of a senior manager's or professional's decision-making process, it should also be able to perform (in its own way) the job the manager is supposed to do. This closely resembles project management perspectives. Here are some hints on needed methodology:

1. Commit one's self to clearly defined goals on mapping time and budget.

2. Schedule a few minutes at the beginning of an operating time period (day, week, month) to establish a program of action.

3. Set goals and priorities.

4. The next step, once the priority list is prepared, is to begin.

Rubinstein, the late pianist, once said the hardest part of practicing was sitting down at the keyboard. Humans have a resistance to changing the frame of reference prevailing at a given moment. Nothing of this type is known presently of expert systems.

Another human weakness expert systems may avoid is the fact that the most notorious of all time wasters is preoccupation: When attention is engaged in something other than that to which attention should be given.

Like the great professional or successful executive the expert system should be able to filter out the irrelevant. It should do this by defining an interactive exchange, a memo being written, or a report.

Just like the masters of finance and industry, instead of having "subordinates" bringing in problems, the expert system should have them bring the answers— the system itself asking the questions. In a way quite similar to successful project managers, the system should not sit on projects, but pass them off to others thereby giving them authority and responsibility to do them fast and well.

Without any doubt, the expert system should review and control objectives. In a dynamic environment, goals and other means it may have put to work, are constantly changing.

7.4   LEARNING FROM EXPERIENCE

Expert systems can learn from experience, like people do, to further develop their skills. At the current state of the art, they can do so within concrete areas of implementation rather than in a general, very broad sense.

This resembles closely enough human talent. In nearly every business, there is in existence much of the information and knowhow required for a thorough review of a product, a group of products, or a function—and such review can be of tremendous control advantage. But in a human organization such information is often in a scattered condition that it is useless for "crisis" reviews.

Information is useless until it has been specially gathered and analyzed. Sometimes the "crisis" in the financial or industrial concern is, or appears to be, so imminent that top managers cannot wait for detailed and specific analyses. So they do the next best thing by issuing arbitrary orders, depending more upon their highly developed "feel" than on the facts they do not have at hand.

At other times, again in human organizations, management though hard pressed by an impending crisis, waits from 30 to 60 days to reach a definite decision. In certain cases, this is a "necessary" waiting period while special analyses are prepared in an effort to arrive at a more intelligent course of action. In these cases the thirty-to-sixty day delay may well prove to be the worst of the two evils for the business, the other being the forementioned alternative of the "feel" method.

While delays characterize a human organization, computer-based expert systems do not have to abide by such constraints. If properly instructed, they can react fast by using the programs developed for this purpose and reaching at electronic speeds the contents of large databases.

This, too, is a major difference from what takes place in human organizations. While sometimes ideas and plans of action must be seasoned in the top executive's mind until a definite decision is reached, in most cases delays in decision making come about as a result of incomplete or overflowing information, and a general overload of work.

Interestingly enough, in human organizations much of this work does not belong to the shoulders of the person it is falling upon. Many executives find themselves spending most of their time dealing with day-to-day detail. The ultimately important is being sacrificed in favor of the immediately urgent.

Parkinson's Law of Triviality states that the time spent on any item of the agenda is in inverse proportion to the value involved. This could hardly be associated with anything bearing the name of "rational decision making" fitting within the overall framework of a successful management strategy. Yet, it is very often practiced. Sometimes top executives in the company do not have the time to realize how important it is to move far enough away from the day-to-day activities to view company matters in perspective.

Expert systems could be programmed to do just that, by following a formal doctrine of decision: the estimate of the situation. It can be described as a series of five formal steps:

1. Determination of the mission

2. Description of the situation and courses of action

3. Analysis of opposing courses of action

4. Comparison of own courses of action

5. The decision

Experience building such systems is necessary to reflect the role of human judgment. Decisions will always be made for a future which continues to be unpredictable, therefore, they always entail risks.

But it is just as true that expert systems can benefit from the rules in aiding decision making many organizations have adopted. This was done on the belief that rather simple steps can greatly improve the performance of a decision maker. These steps are basically four:

1. *Define the Problem**. What kind of problem is it? When do we have to solve it? Why do we solve it? What will solving it cost?

2. *Define the Expectations.* What do we want to gain by solving it? Short and longer-term goals? Resources? Timetables?

3. *Develop Alternative Solutions.* How many feasible solutions exist? What is their relative profitability? Which of several solutions offers the "best" possibilities?

4. *Know what to do with the decision after it is reached.* The reason humans thought it might be possible to help problem solution by organizing is found in that the approach to problem solution is really an attitude of mind. It might be defined as bringing the research point of view to bear on all phases of the business. This involves assembling the facts, analyzing where the facts appear to point, and having the courage to follow the trail indicated even if it leads into unfamiliar and unexplored territory.

This approach to decision making is never satisfied with things as they are. It assumes that everything and anything—whether it be product, process, method, procedure, or social or human relations—can be improved. Perhaps, the best single term that can be used to describe this attitude, is the *inquiring mind.*

This is the type of mind which is always seeking for new ways, rejecting the commonly used thought that something is "good enough" because it happened in the past. Expert systems can ease the task of decision making, and get better results, by recognizing that every decision has two distinct ingredients. The first and most obvious is quality. To assure the best decision, its maker should assemble facts, consult other experts, and dig deeply into his own fund of knowledge and experience. The second ingredient is acceptance. The effective-

---

*And remember: describing is *not* defining.

ness of a decision depends on the degree to which those responsible for carrying it out like it, believe in it, and execute it.

High-quality, low-acceptance requirements characterize problems in which quality is the important ingredient and the need for acceptance is relatively low. Such problems should be solved best by expert systems: Acceptance should be considered only after the quality of the decision has been assured.

High-acceptance, low-quality requirements are native to problems in which poor acceptance can cause a decision to ultimately fail: The judgement of quality is influenced by differences in position, experience, attitudes, value systems, and other factors. For some years to come, in this particular area, human decision makers will do better than expert systems.

7.5   BORDERING ON IMAGINATION

An important subject within the aspects of executive decision making under discussion is *creativity*. Simply stated, creativity put into a decision, by its maker, can help both its acceptance and its quality. Studies thus far conducted to determine what makes a person creative point to five principal characteristics:

1. Problem sensitivity
2. Idea fluency
3. Originality
4. Solution flexibility
5. Synthetic ability

Problem sensitivity is basically the quality enabling the executive to recognize that a problem exists; or to be able to cut through misunderstanding, misconception, lack of facts, or other obscuring handicaps, and identify the real problem.

We don't know yet how to integrate such qualities into a computer-based expert system, but we do appreciate that a good way to improve problem sensitivity is to keep in mind that nothing is ever as well done as it could be. Every man-made article, every business operation, every human relations technique can be improved and someday will be.

The chance of the expert system comes in by way of the fact that in every situation an executive encounters, no matter how many times he has met and handled it before, an opportunity exists to find a better way. If we learn to recognize these problems as challenges to our own creative effort, we will be half-way to finding creative solutions to such opportunities.

Idea fluency presents challenges and problems. Ideas can range in value from the completely new abstract mathematical theory down to a way to save a dollar a day in the drug store. In practical, everyday business problem solving, complete newness or pure originality, is usually not what is needed. In fact, it may not even be wanted.

The *originality* required of the executive is more likely to be that of finding new ways to vary existing conditions, or new ways to adapt existing ideas to new conditions, or a new modification of something that will fit in an existing condition.

An expert system might be ingenious along that path. The difference between a great executive and an ordinary one is often the ability to produce original variations to meet existing conditions. The relative attribute of originality can also be developed, or at least simulated, to the point where it meets the requirements of successful business operation.

In fact, experiments have demonstrated that all five forementioned characteristics can be acquired or developed, to some degree, in any individual. This does not mean, of course, that a person who rates low in using his imaginative faculties can suddenly be turned into a creative master. But he/she can, through application, learn to do more with what he/she has.

If naturally creative people can, through experience, learn to raise their already high creative output even higher—an expert system can be taught in following this road. The quality of *creative flexibility* is largely that of being willing to consider a wide variety of approaches to a problem. This, in turn, is largely a matter of attitude—or of euristic programming.

Rather than obstinately freezing into one particular idea, or a single approach to a problem, the flexible person, or expert system, starts out by remembering that if one solution won't work, he can always approach the problem from another angle. Wherever there is a solution, there is an alternative. The creative person just plain expects to solve the problem, no matter how many failures temporarily delay the solution.

Synthetic ability is necessary in order to arrive at a concretization of the solution and to pull together all its component elements. Executives have often been helped in this step by setting a deadline, to push oneself mentally.

If the executive really wants to get himself mentally involved in meeting that deadline, it might be a good idea to let others know about his self-imposed deadline and his intent to meet it. Expert systems can follow quite a similar road, and they can do so much more effectively than humans.

Controlling through milestones the time, the money, and the people depending on an executive position is a wise mode. The same is true of handling each project as if one's professional status depended on it.

The executive can define his problem in an effective manner if he develops the ability of locating the *salient factor*. The salient factor is the one element in the situation which has to be changed, moved, or removed before anything else can be done. By isolating this strategic element we usually manage to break through the surface of the symptoms and grasp the real problem.

In order to locate the salient factor, the executive must first set the *criteria*. Such criteria are usually a function of his/her *objective*. The same criteria will also be used later in the evaluation of alternative plans. They must include:

- *Risk level*—How much risk the organization is able or willing to undertake.

• *Timing*—A decision should not be made before it can be effective, nor should it be postponed after it has reached a point where it has become timely.

• *Sensitivity to change*—How much change a given decision will withstand before becoming obsolete?

What makes the decision-making process rise above a given risk level is the number of pitfalls which it includes: searching for an answer to the wrong problem, taking more risk than one can afford, making the decision at the wrong time, or selecting a course of action one has not the means to follow.

Failures in decision making can be attributed to one or more of the following reasons: trying to avoid thinking, taking the easiest path, failing to organize the decision process, poorly distributing the time spent on making the decision between defining the problem and finding the answer, tending to look upon a decision as a problem rather than as an opportunity, tending to settle for the solution which has the lowest cost, even though it promises the lowest gain and entails the greatest risk.

Expert systems can be programmed to avoid such pitfalls. This does not mean that we know already how to build computer-run processes with imagination—though that day may eventually come.

One of the top criteria of intelligence will eventually be the ability to reflect on changes in environmental situations which account for some of the worst failures in financial and industrial decisions. Political change, economic change, wars that have not been forecasted, or have not been properly anticipated in timing, may prove to be decisive on the success or the failure of a decision, and sometimes of the executive who made it.

Each action chosen needs to be identified with a distribution of potential outcomes. Only one will materialize, and that one cannot be foreseen with certainty. Therefore, making a decision is choosing the action whose potential outcome distribution is preferable. The choice is necessarily a function of the skill of the decision maker, and of the assistance he can get at the proper moment.

7.6 MANAGEMENT BY OBJECTIVES

A highly potent, proven approach for prompting both the efficiency and effectiveness of executive positions is management by objectives. This is an area where expert systems may excel.

Few managers would argue against the wisdom and desirability of planning out their operations by establishing objectives, scheduling the activities necessary for their accomplishment. However, considerable experience indicates that too many managers jump right into the writing of specifications. The results are usually weak specifications which cause little if anything to happen.

The fact that the efficient manager needs an organized, step-by-step thought process leads to the consideration that expert systems call for nothing else. Such organized approach helps in two ways:

1. First, in the selection of objectives.

2. Second, in determining at what level of achievement or performance the objectives should be set.

Table 7.2 presents a step-by-step process which has been used by countless people to better manage their operations. Expert systems ask for nothing more. The objectives which will ultimately be written will flow from and be a part of the outlined process—provided those in charge of it understand what the responsibility really is.

TABLE 7.2 A step by step process to manage new systems put in operation

1. Clarify the objectives.
2. What do we require?
3. What do they expect to accomplish?
4. How can we select key results areas?
5. Outline the capabilities to be secured by the system.
6. Write down the objectives: how, what, and when.
7. Plan how the objectives will be achieved.
8. Identify the training requirement of the human resources.
9. Train and install.
10. Demonstrate the usage, help in the implementation.
11. Maintain and upgrade.

In this particular count, expert systems may excel on their human counterparts. A considerable body of research indicates a significant percent difference of understanding between a subordinate and his/her boss, as to what is expected of the subordinate.

Also with procedural systems, while we often talk of objectives we tend to forget the long-range nature of many projects. Installing an integrated computer-based system, for example, is no immediate undertaking.

Expert systems can better remember the long lead time required for delivery of a working ensemble. They can keep track of the rapid rate of obsolescence of equipment, account for the scarcity of personnel in many specialties, and never forget about costs—as sizable sums are involved whenever high technology dominates the scene.

Like the typical manager, an expert system would receive major guidance by the user departments through their: objectives, plans, and specific references as to what is required. Both may secure this information through copies of the objectives and plans or by other means of consultation. Such inputs, combined with the inputs from different levels, constitute the base which permits to proceed with the specifications.

The expert system stands a good chance to find itself ahead of its human counterpart in performing these chores: If inconsistent determinations were made in different departments for situations which are very much alike, the affected population would lose confidence in and react against the decision makers and the decision-making process as a whole. But a computer-based approach should be able to weed out such discrepancies.

Marshall Foch once said that there can be no collective harmony in the active sense in any organization unless each and every one concerned knows what the purpose is. Purpose being consistent and of unique nature, translates itself into a doctrine of unity, involving a single spirit which motivates continuous and coordinated efficiency.

The military attains this uniform of doctrine through identity of training and strict, well-defined, inflexible regulations. Expert systems can behave in a similar manner. Like certain industries striving toward uniformity in more or less routine-type affairs, expert systems ingredients can be identified as follows:

1. Transcribe knowledge into written procedural form
2. Develop extensive instructional outlines
3. Make regular performance evaluations
4. Institute case references
5. Direct and coordinated reviews
6. Implement administrative analysis and pilot or stipulated cases
7. Use interpretation units and program consultancy

Some of these ingredients are not new. Transcribing knowledge in the form or organizational policies, procedural interpretations, forms manuals, and precedence are the basic ways for achieving uniformity in routine judgmental situations. With expert systems, the creation of guiding principles or standards for specific end products may include programmatic interfaces.

To be of value, these programmatic interfaces should acquire operational acceptance prior to implementation.

Performance evaluation techniques are a good example of routine-type decisions where expert systems can be of help, particularly so if remedial action is integrated. Management should always keep under perspective the recommended corrective action.

Through expert systems and other coordination means a company can achieve sharing of the evaluation skill available within itself. With typical managerial approaches, at the level of routine-type decision making, oneness of thinking can be created through the permeation of knowledge. An expert system will assure this function.

Quite to the contrary, uniformity of decisions is neither applicable nor desirable in non-routine matters confronting the enterprise. Such is the case in one of the most intricate decision-making areas: that of diagnosis and of subsequent pre-scription. Even when a diagnosis is correctly done, the most difficult part of a consultation can still be present.

Let's recapitulate. Though exceptions exist in every process and imaginative steps are then a basic requirement, many executive situations are amenage to management through expert systems. In these, a key role is played by a well-studied transition to help in reflecting the way decisions are elaborated and documented.

Expert systems development will be the more successful if the business trend is studied in detail in terms of the critical variables: for instance, growth in transaction volumes, geographic distribution of operations, shifts in demand, increasingly complex product line reporting needs.

Developments have occurred along this line during a period when important advances were being made in data processing technology and its application. These advances include:

1. Increased computer system usage

2. Switch to interactive solutions

3. Computer vision systems

4. Advanced use of data communications facilities

5. Distributed data processing

6. Network applications

7. Distributed databases

8. Management oriented applications systems

The development of expert systems is in no way inhibited by the fact that there have been management information systems failures in the early 1970s. At that time, we tried to work on the total mass of information, without the appropriate means. However, since then, with the creation of large databases, we have controlled significant amounts of information.

At the time there was no database concept as such. Today there is in terms of organization, update, and DBMS. Again at that time there was no immediate update of management information—with the result that data were obsolete and of little use to managers. Through online solutions, there is an interactive update. Data is actual.

Twelve to fifteen years ago mathematical models were taught to the manager and the experience was a dismal failure. Today the overriding demand is for user-friendly solutions which have made the success of the spreadsheet. In "what if" situations, the model is transparent.

In the 1970s there was a split, almost watertight, between data, text, and voice. This partition is about to disappear through data, text, image, and voice integration.

Finally, at the time man-information communication was tedious and demanding specialist skill—hence, it discouraged management. Today, through menu selection and other interactive approaches, we have a user-friendly environment. Tomorrow, this will become still easier through voice detection and answerback.

Chapter 8

PREPARING FOR DECISION SUPPORT

8.1 INTRODUCTION

The primary object of computer-based decision support systems (DSS) is to develop *aids* to management decision. Such aids involve visualization (graphics presentation, and algorithmic expressions) which, however, are transparent to the user.

The projection and implementation of decision support systems imposes prerequisites. For instance, we must assure compatibility between system structure, communications solutions, supported processes, databases, and relationships between the information elements in these databases.

One of the major benefits of a DSS software is that of giving managers and professionals considerable powers of calculation without requiring programming experience. With a good interactive DSS, the user can incorporate significant volumes of information in a given model, and then change a condition of the model and redo the entire evaluation, analysis, and presentation.

Though the investment in user training can be higher than for simple functions like electronic mail, the benefits can also be much greater. But model choice has to be validated through both pilot runs (prototyping) and postimplementation.

8.2 LOOKING AT DSS IN A PRODUCTIVE WAY

While algorithmic approaches are the skeleton of a decision support system, this skeleton is worth nothing without muscles: the supporting text and data. Data represents 80% of the development effort and is the key ingredient in the results obtained.

The algorithms are the tools for analysing the data once the information has been received. This is true of financial modeling tools as it is of graphical representation aids that can assist the professional in working with the data.

The facilities that can be used directly by professionals and management will be a direct reflection of the accuracy such data represent. This implies that throughout the design process we must plan for integrity, and this calls for error reduction, journaling, and security/protection.

Effective use of DSS technology also involves a good deal of support media for instantaneous response. The whole difference made in the past between batch and realtime comes under perspective, but with some added requirements. We must:

1. Design for *directional*, rather than positional data.

2. Plan for an integration of goals between decision makers and information providers.

3. Project (in terms of computer services) aims capable of responding to the information needs of the users at each level.

4. Give every user the possibility for experimentation (*as if* tests).

5. Place emphasis on graphics, using color for exceptions and keeping an order to reflect specific functionality.

Strategic objectives must be considered and the overall structure of a DSS should fit the course of management practice. This usually takes the form of critical evaluations and mission goals. Figure 8.1 identifies an often followed path in senior management decisions.

The experience from recent projects underlines that new concepts are necessary both at the end user and at the specialist level. The whole image that analysts have of a computer-based system must be changed. Let me take the management of installment loans as an example.

Some months ago, a given financial institution decided to computerize its installment loans operations which had been done mainly on a manual basis. Yet, while an online system was clearly specified, the analyst responsible for this work thought in batch terms and projected a basically batch system—with the end result that the project had to be done once all over again.

This cost time and money, but it also offered the possibility of doing some interesting comparisons. Tests demonstrated that the advantages derived from the WS-based interactive solution exceeded by nearly an order of magnitude the results of online to mainframe, batch looking procedures—in terms of derived benefits. Any organization should, therefore, ask itself: "Are we going to keep on paying 10 times the cost of personnel at the present state of very costly human services, to keep the old images in place?"

Both easily evident and hidden costs must be considered. If the professional employee, to whom the system is addressed, is obliged (or even permitted) to keep the large paper files while given the new facility—the result will be an increase in cost and in the usage of his time.

It is unwise to implement a DSS without the proverbial long, hard look on the operations to be supported, the effect on the end user, the training requirements, the abandoning of old, obsolete practices, and the transition routines.

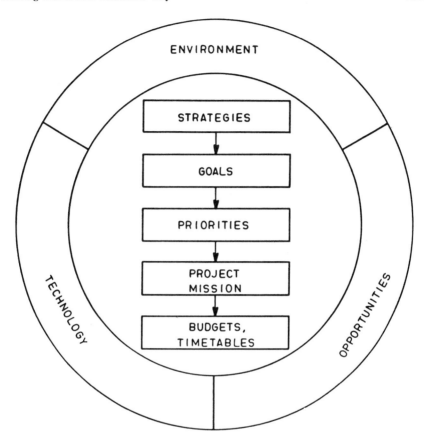

Figure 8.1 Projecting a Strategic Information System.

Every step has to be planned, and its aftermath preevaluated prior to having it finalized.

The work on decision support systems does not end with the implementation of interactivity. There are plenty of other things which need to be done in developing the needed understructure. A recommended development of a computerized, interactive system would:

1. Accommodate external as well as internal data sources.

2. Provide for narrative and statistical sets of data able to reside in the database under a catalog of available information.

3. Have a time series-oriented data storage facility, and make use of inference algorithms.

4. Assure flexibility of analysis including spontaneous as well as scheduled types of investigation.

5. Be open to graphics, preferably color.

6. Avail an output facility capable of being handled at the WS level preferably through video displays.

7. Not only be interactive but also capable of integrating with other functions handled at the WS level.

8. Enable the user to communicate directly and quickly with data and to evaluate alternatives, making full use of the system's analytical aids.

9. Provide features which allow a decision maker to pose and repose questions considered crucial for decisions.

10. Incorporate interfaces making the system easy to use.

11. Be adaptable to a changeable environment through an accommodating system architecture.

12. Anticipate an evolutionary development cycle. A modular approach suggests itself.

The whole project must be oriented toward the end user and assure an able, timely, cost/effective response to the user needs. User modules consist of a series of related decision options and decision processes. Both abide to rules and procedures.

The environment should be user friendly. In this connection, *menus* comprise an important aspect of the design phase. They make possible sets of choices developed to visualize the alternatives to be analyzed for the decision.

Based on a utilitarian design concept, menus enable the decision maker to select from a table of earlier choices or to develop new approaches for examining the relevant factors of the decision. This has been, for instance, the design principle of Videotex.

Evolutionary developments should be foreseen in all main references: data-basing, datacomm, data processing, end user functions. Design considerations should surely account for the fact that the needed information for decision making grows more rapidly than most data systems can effectively provide—unless they are projected with this goal in mind.

8.3   CHOOSING CANDIDATES FOR DSS

It is unwise to bet the DSS project on a long shot. Rather than doing high-risk innovations the "big way," serious users will be well-advised to set up a small pilot project. This allows one to take chances on promising innovations without courting disaster.

The introduction to the utilization of decision support systems as a matter of course can be made that much more interesting for the user (and profitable for the organization) if the subjects are well chosen, easy to use, and of substantial aid to decision.

This is the case with ratios such as:

The Acid Test (**Current Assets to Current Debt**).   Current Assets are the sum of cash, notes and accounts receivable (less reserves for bad debt), advances on merchandise, merchandise inventories, and listed securities not in excess of market value. Current Debt is the total of all liabilities falling due within one year. This is the test of solvency.

Collection Period.   Annual net sales rates divided by 365 to obtain the average daily credit sales. Then, the average daily credit sales are divided into notes and accounts receivable, including any discounted. This ratio is helpful in analyzing the collectibility of receivables.

Current Debt of Inventory.   Dividing the current debt by inventory yields; another indication of the extent to which the business relies on the funds from disposal of unsold inventories to meet its debts.

Current Debt to Tangible Net Worth.   Derived by dividing current debt by tangible net worth. Ordinarily, a business begins to pile up trouble when this relationship exceeds 80 percent.

Depreciation.   The cost to a firm of using up its machinery and other assets resulting from wear, decay, or obsolescence.

Discount Rate.   The interest that Federal Reserve banks charge on funds borrowed by commercial banks that are members of the Federal Reserve System.

Exchange Rate.   The amount of one currency that is traded for another currency by bankers and dealers.

Fixed Assets to Tangible Net Worth.   Fixed assets are divided by Tangible Net Worth. Fixed assets represent depreciated book values of building, leasehold improvements, machinery, furniture, fixtures, tools, and other physical equipment, plus land, if any, and valued at cost or appraised market value. Ordinarily, this relationship should not exceed 90 percent for a manufacturer and 75 percent for a wholesaler or retailer.

Funded Debts to Net Working Capital.   Funded debts are all long-term obligations, as represented by mortgages, bonds, debentures, term loans, serial notes, and other types of liabilities maturing more than one year from statement date. Financial analysts tend to compare funded debts with net working capital in determining whether or not long term debts are in proper proportion. Ordinarily, this relationship should not exceed 100 percent.

Inflation.   A protracted rise in the general price level. Over any significant period of time, this is associated with an increase in the supply of money.

Interest.   The amount a borrower pays in return for use of a lender's money.

Inventories.   The supply of raw materials and unsold goods that a business keeps on hand to meet production and customer needs as they arise.

Inventory to Net Working Capital.   Merchandise inventory is divided by net working capital. This is an additional measure of inventory balance. Normally, the relationships should not exceed 80 percent.

Liabilities.   Money a business, government, or individual owes to others. Includes amounts owed for mortgages, supplies, wages, salaries, accrued taxes, and other debts.

Net Profit on Net Sales.   Obtained by dividing net earnings of the business, after taxes, by net sales (the dollar volume less returns, allowances, and cash discounts). This is an important yardstick. It helps measure profitability and should be related to the ratio which follows.

Net Profits on Net Working Capital.   This represents the excess of current assets over current debt. As a margin, it identifies the cushion available to the business for carrying inventories and receivables, and for financing day-to-day operations. The ratio is obtained by dividing net profits, after taxes, by net working capital.

Net Profits on Tangible Net Worth.   Tangible net worth is the equity of stockholders in the business, as obtained by subtracting total liabilities from total assets, and then deducting intangibles. The ratio is obtained by dividing net profits, after taxes, by tangible net worth. The tendency is to look increasingly to this ratio as a final criterion of profitability. Generally, a relationship of at least 10 percent is regarded as a desirable objective for providing dividends plus funds for future growth.

Net Sales to Inventory.   Obtained by dividing annual net sales by merchandise inventory as carried on the balance sheet. This quotient does not yield an actual physical turnover. It provides a yardstick for comparing stock-to-sales ratios of one concern with another or with those for the industry.

Net Sales to Tangible Net Worth.   Net sales are divided by tangible net worth. This gives a measure of relative turnover of invested capital.

Net Sales to Net Working Capital.   Net sales are divided by net working capital; it provides a guide as to the extent the company is turning its working capital and the margin of operating funds.

Price-earnings ratio.   The market price of a stock divided by earnings per share for the previous 12 months. A stock selling at $20 and earning $2 per share has a price-earnings ratio of 10.

Prime rate.   The interest that banks charge on business loans to their strongest customers.

Productivity.   The relationship between "output," or the quantity of goods and services produced, and "input," or the amount of labor, material, and capital expended to produce those goods and services. Usually measured in terms of output per man-hour.

Profit.   Money remaining after all costs of operating a business are paid.

Return on Investment.   The amount of pretax profit from an investment stated as a percentage of the original outlay or purchase price.

Seasonal Adjustment.   An adjustment in business statistics to take account of more or less regular movements that recur year after year, such as the increase in retail sales associated with the Christmas season.

Total Debt to Tangible Net Worth.   Obtained by dividing total current plus long-term debts by tangible net worth. When this relationship exceeds 100 percent, the equity of creditors in the assets of the business exceeds that of owners.

Yield.   The annual return on an investment expressed as a percentage of the investment's cost or current market value.

This is one of the issues management can understand without problems. Yield must not only be presented in quantitative terms for each item in the inventory (whether it is goods, stock, bonds, or any other item) but also the presentation must be made in a way that the trees do not hide the forest.

If the presentation is in a tabular form, three significant digits should suffice. For management accounting (as contrasted to General Ledger) it is counterproductive to focus on dollars and cents—losing from sight possible differences in millions.

Still better for management accounting purposes is to present the results in a graphic form. Different hypotheses tested through a DSS model can each be given a specific color, making more visible the interpretation of the results.

In conclusion, the DSS references we have just considered help provide management with services oriented toward security, liquidity, and critical evaluation. Results can point to effective control action and/or an investment strategy tailored specifically to the company's requirements.

Three critical issues can be addressed through DSS, among other subjects. First is the *preservation of capital*, which makes feasible the professional management of cash with liquidity and low-risk objectives. Second is *the growth of capital*, which is measured, for example, by estimated changes in demand for a given security or in terms of investment opportunities. Third is *the control of risk*, through the steady, objective evaluation of quality and liquidity of chosen investments, portfolio and currency diversification, and forecasts of market volatility. DSS programs can also consider current investment trends, focusing on those which offer the greatest stability and potential for earnings.

8.4 RETRIEVING THE INFOPAGES

The results of decision support experimentation must be presented in a comprehensive form, with particular emphasis on the ability to pinpoint variations. Furthermore, user access to these infopages should be linear and properly organized—preferably through *menus*.

Menus are a common technique for permitting users to communicate with their WS and the microfiles which it supports. A menu is a list of options—creating the route to the desired *infopage* which contains the requested information.

The user needs only to indicate which option he/she desires by pressing the corresponding numerical key. The sequence of options made in successive menus creates a routing path whose mechanics are transparent to the user. It is also a recommended practice to build by bypassing the menu sequence, calling the required infopage in a *direct* way.

Typically, a menu will offer up to 10 options, but it is a good policy to use less than that, leaving room for expansion. Since options (the routing pages) and the final infopage are presented in page form, it is wise to organize the information in the database in a page format—each page being properly identified.

Page numbering must be done in a way which is easy to access, offers more than one database entry possibility, is open to expansion, and can handle a variety of applications without changing page form.

Figure 8.2 demonstrates a two-way organization of the menu sequence contained in the database. The infopage will be found at the end of the tree and may contain tabular data, text, graphics, or other forms of presentation.

For infopage numbering purposes, a suggested numbering system is:

Basic Code		Suffix		Origin
⟨bc⟩		⟨s⟩		⟨o⟩
xxxx	—	xx	—	xx

Such a system can support a database entry which includes menu selection capabilities, as well as direct access by page number and the use of keywords. Mnemonic keywords can be converted into ⟨bc⟩⟨s⟩⟨o⟩ through table lookup.

Quite importantly, the fundamental understructure of the foregoing system is invariant whether we talk of applications in finance, in industry, in government, or in other sectors.

The basic code can be a running number; the suffix may reflect functional characteristics; the origin identifies the source. With interactive applications, it is quite important to show:

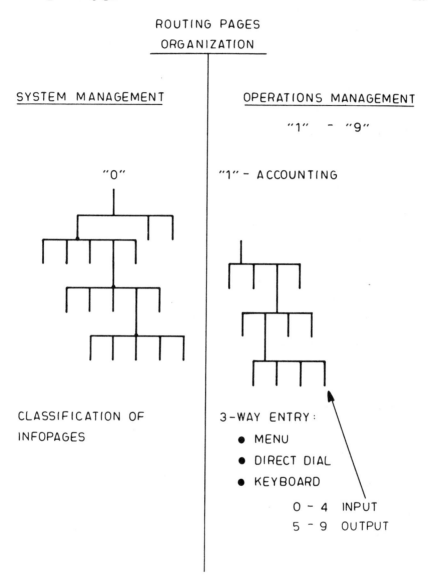

Figure 8.2 Routing Pages.

- the name of the originator for the document or message
- the originator's organization
- the origination date
- indicators such as whether it is *personal* or *confidential*, or that a *return receipt* has been requested

By depressing a program function key or similar media the user should be able to initiate selection of a specific item from a queue or general database storage. The process should be tutorial in that the actions and corresponding keys appropriate to the task are shown on the screen.

When the user has finished viewing an infopage, he can select one of several options related to its disposition. Depending on job requirements (and supported features), he could place the document in a suspense or *hold* status for action at a later date; alternatively he may forward it to other users, or for archiving.

All this clearly emphasizes the need for a *systems understructure*. A part of the able choice of ratios, tabular presentation, and graphics forms to aid management decision, success or failure of interactive solutions will necessarily involve:

- properly designed workstations
- the needed software support
- algorithmic approaches (processes)
- databases and database management
- user interfaces and related supporting gear
- costs and cost/effectiveness

The systems study, though necessary, is not enough. We need procedures for:

1. *Systems management*
   - Creation
   - Classification
   - Maintenance.

This involves both routing pages and infopages. They must be handled in a standardized, homogeneous manner no matter which may be the application.

2. *Operations management*
   - The use of the input/output infopages in relation with the database content
   - The way the interactive work should be done day-to-day for operational reasons
   - The instruction of the end user so that he can most effectively work with the interactive system

The end user should not only be trained on how to approach and operate the interactive system, but also he should be a major participant in systems design. The end user is the person most concerned with the results and he must contribute to the study which will help assure these results.

Users are sensitive to both the physical and to the logical environment. To give an able answer to logical questions, we have to care about *software support* for

- Computing
- Formatting
- Communicating
- Handling the aspects

Software facilities should be transparent to the end user—though the user's system specialist should not only know and understand what they mean but also be able to request the needed supports by the manufacturer. Such supports are part and parcel of an optimal vendor selection.

Software will be usually supplied at two different levels: a set of primitives and complete packages. The final choice will depend on objectives and the available skill to do the implementation job. Yet, whatever the case may be, the user is well advised to keep out of writing the system's software. Office automation is the first computer field where packages have achieved general acceptance.

Another point of emphasis is the design and use of *the database*; both its organization/structuring and the security problems it presents. The organization of information contained in the database is of crucial importance as such arrangement will prove useful for a variety of retrievals, interactions, performance characteristics, and so on. General database technology, for instance, can be improved through the process of constructing an interactive system.

Take, as an example, the sales manager's job of continuing evaluation of field people, markets, and the impact of competition. Usually, this information is hidden in the company's own voluminous alphanumeric files while other references must be digested from public information, market intelligence, and the like.

Typically, the people who gather this information are not trained to give the manager all that he needs. With batch systems, the time it takes to accumulate and analyze these data compounds the difficulty—and, as already stated, many online systems resemble batch processing.

Contrast this with the process of entering data on salesmen, order books, customers, and territories against an appropriate geographic base file and using standard programs to compare the relevant variables. The understructure of the interactive environment starts at this very point, but needs to be enforced through proper design considerations.

8.5 KEEPING THE OPTIONS OPEN

While algorithmic solutions are, as stated, the skeleton of DSS and data is its muscles, visualization can help the presentation and appreciation of critical data. Color can be used, for example, to show typical ratio figures, upper and lower quartiles, and internal data or the experience of firms in the top and bottom halves of a sample.

A leading bank has found considerable benefit in making interactively available to its management financial ratios exceeding the net maturity indicated by selling terms, comparing collection period, and evaluating allowances for possible variations in selling or collection time. Trends, tendencies, changes, and variations are demonstrated in an action-oriented manner through the use of computer graphics.

To simplify development of forecasts and estimates of interrelationships between elements from time series, an industrial concern has chosen supporting software providing statistical analysis techniques accessible interactively. (The following are among the employed standard routines we distinguish: simple regression, moving average, multiple regression, seasonal analysis, and exponential smoothing.)

Usage ranges from market forecasting to inventory control procedures. The responsible executive enters all important work parameters and has the results presented in graphic form and color. The mechanics are simple: An interface program makes a logical test and then sets the color code. But before color codes are assigned, a special action is required to ascertain whether the item under consideration is in stock below a specified minimum number:

• For some products, the number of "weeks' cover" in the column is examined and compared to standard.

• For other products, the same is done for the number of "days' cover" column (to check if it is below a minimum number of days stock).

• For all products, the stock figures on each page are examined for negative values.

These are the basic prerequisites for exception reporting and its implementation through color presentation. The practice is good: Any firm can use a similar approach to color both the tables and the graphs which it needs to develop for management reporting.

Since they are organized in infopages and are accessible through menu selection, the results of DSS evaluations made in one site can be easily transmitted to another in a standard format. This is a practice we are successfully following with videotex throughout business and industry.

Keeping the options open helps in enlarging the domain of implementation. *Time management* through schedule calenders and calendar update is an example easily coming to mind. *Document handling* is another.

Menu presentation, of which we already spoke, is just as applicable with document handling procedures as it is with DSS infopages. In fact, the two facilities nicely complement one another.

A *pass for action*, computer based, service gives the user the ability to pass on a document to others. When this option is selected, the user can be presented with a list of names of people with whom the most frequent communications occur.

The user selects any of the names listed, or can key in other names at the bottom of the screen. After the selections are made, the system displays a list of pass-for-action messages for each name selected such as: for your information, please see me on this item, and give me your comments. In addition to these prestored messages, the user can key in a message of his choice. He can enter an optional suspense date which will be added to the message, and so on.

A reminder can be automatically inserted into the sender's hold queue showing to whom he passed the document, the sequence date (if any), and the message. The user can also retrieve, at his ease, any background page with text, data, and graphs he considers important.

Time management is by excellence an application which supplements and integrates with DSS. Personal schedules can be carried in system storage in a queue similar to that used for electronic mail. Each entry in the queue may, for instance, correspond to a day, with appropriate identification in the queue entry.

To examine the schedule, the manager selects a day by depressing a key. The schedule for that day is displayed, and may be paged forward or backward by key depression, or other actuator, just as documents being viewed can be paged.

The format of the day's schedule should be kept simple, containing the hours committed, the individuals involved, and a brief statement of the purpose of the meeting or appointment.

Remaining open periods can easily be indicated. Since the statement or purpose portion of the schedule is free text, notations to remind the principal of departure and travel time, airline bookings, reservations, and so on can be included, with no limitations on the span of days that may be carried by the system.

Most available software makes it feasible to define the workday to the system in a way which suits the individual user. They also keep open slots in the schedule that match the work habits of the individual manager or professional.

Entries into the schedule are made interactively, with the system soliciting the required information, and the user responding with minimal key strokes. Time schedules may be scanned by the system and a display of only the open slots presented, if this is the requested action.

Software will also see to it that appointments are easily cancelled, and conflicts in making them are flashed out, necessitating the rescheduling of affected appointments. Such conflicts are called to the user's attention, and provision is made to reschedule the original appointments to other open time periods without reentry of the original information.

Time scheduling systems also provide for group scheduling capability. This helps match the open time available for each person selected, presents the best time when all will be available, and indicates the persons causing a conflict. They also allow the user to enter reminders of nonscheduled events or actions.

All this is written to document that a Decision Support System is nothing monolithic and it is not only mathematical. As experience is gained with its usage, more requirements are added in terms of services. These tend to merge different formerly independent facilities into one well-coordinated aggregate.

Chapter 9

USING THE SPREADSHEET

9.1 INTRODUCTION

Spreadsheets are the simplest programming tools among the Fourth Generation Languages. Originally designed as applications packages to enable the use of personal computers in complex financial projections and the displaying of results, they developed into user-friendly tools for machine programming.

Visicalc, the original spreadsheet, was designed in 1979 by Daniel Bricklin for operation on an Apple personal computer. Since then, spreadsheets have gotten considerably more sophisticated. Graphics capability has been provided particularly for computer-generated pie charts, bar charts and other management type displays. New additions, in an integrated applications environment have been: word processing, electronic mail, and facilities for transmitting and receiving data over telephone lines. These are examples of developments in *integrated software*; that is, programs that combine many of the most popular applications for a personal computer—beyond the original financial management facilities.

For instance, through integrated software tools, stock data can be retrieved by telephone and analyzed on a spreadsheet. Its embedded program facilities will enable the user to manipulate long rows of interrelated figures for financial projections. The preparation of economic proposals, portfolio handling, the management of personal treasury, and consolidation of accounts are other examples served by a spreadsheet in an able manner.

The associated graphics and word processing capability helps in the easy-to-follow presentation of the results in an exception-reporting style. Such results can be put instantly by the computer into letter or report form alongside charts with color display of important issues. Color helps in further highlighting a trend.

New releases of this type of user friendly software make use of windows; by splitting the screen into segments the user can view at the same time. What's more, the user can move from one to another segment through the press of a button or other man-information communications device attached to the machine.

The object of this chapter is to demonstrate the ease with which the power embedded in a spreadsheet can be employed by a computer. The first sections stress the fundamental operations supported by a spreadsheet; the latter give an applications example.

The software product employed in this example is Microsoft's Multiplan. The personal computer is Olivetti's M20, and the functional keys to which reference is made relate to this machine.

9.2 INITIALIZATION OF THE MACHINE

This presentation is made with the goal of hands-on experience. Purposely, it is written in a matter-of-fact manner. The user can follow it step-by-step in working with the stated spreadsheet and personal computer.

Step 1. Insert the Diskettes

> Right: *Multiplan* Floppy Disc
> Left: Data Floppy for *Personal Archive*

Step 2. Put switch in Position "Power On"

Step 3. Wait until the cursor appears and simultaneously depress the keys "?" and the yellow "COMMAND" followed by "RETURN". Then type:

```
MP
RETURN          ( ← )
```

Step 4. It would then appear as the first empty page of the spreadsheet. The program is ready to work with. All pages are identified with a name. (We will return to this subject.)

9.3 COMMANDS AT THE USER LEVEL

9.3.1 Handling of Cursor

The *cursor* is a pointing device of the form □. It shows as a light window on the screen and takes one character. This is the character on which the user is working at this particular moment—that's why cursor positioning is so important.

The cursor can be moved all over the screen (left, right, up, down), through arrow-labeled keys or other devices such as a mouse. A typical position in which the cursor is returned on the screen is known as the *home*.

With spreadsheets we also use a different cursor: *cell pointer** which represents the intersection of column and row. It is defined by the user and has the number of characters assigned for the column to which it corresponds. Like the cursor, the cell can be moved with the

$$\text{SHIFT} \quad \text{and} \quad \overset{\uparrow}{\underset{\downarrow}{\leftarrow\!-\!\rightarrow}} \quad \text{keys}$$

To return to the initial position use the:

SHIFT   and   HOME   keys.

9.3.2   Explanation of the Keyboard

The keyboard divides into two parts:

- alphanumeric (left)
- only numeric (right)

Each key has the functions written on it: upper and lower case characters. There are also functional (preprogrammed) keys at the upper row of the left part of the typewriter. (Their actuation is conditioned by the use of the yellow COMMAND key.)

The most important keys are:

> RETURN key
> SHIFT (right or left) key
> Yellow COMMAND key.
> blue CRTL (CONTROL) key

The RETURN key's usage follows the input of the commands and is necessary for their execution.

Its usage turns the numeric keys of the first line of the keyboard into *functional keys*. With this key the user selects the line in the bottom of the video form he/ she is presented with.

On the M20, the CONTROL key's functionality is similar to that of the COMMAND key, but it helps in selecting the row in the bottom of the videoform. If the videoform is empty, it doesn't function. The principal functional keys in the functional row (first line on top of the keyboard as discussed in the above paragraphs) are:

---

*Referred to in this text as Pigeonhole.

Argument	Function
CANCEL	Interrupt the command being executed at this moment.
PAGE UP	It moves the infopage (videoform) by one page above the one being viewed (preceding it).
PAGE DOWN	Idem, for the succeeding (following) infopage.
PAGE LEFT	Idem, to the infopage to the left.
PAGE RIGHT	Idem, to the infopage to the right.

The other functional keys, like NEXT WINDOW, NEXT UNLOCKED CELL, etc. are rather special. It is better not to use them in an introductory phase of implementation.

9.3.3   The Meaning of Commands

(The most significant commands are identified with an asterisk: *)

Written	Key	Function
* ALPHA	A	Input of alphanumeric data into the pigeonhole.
* BLANK	B	Empty (reset) the pigeonhole.
COPY	C	Copy content of one pigeonhole into another (same for row or column).

Written	Key	Function
DELETE	D	Empty (reset) all pigeonholes of a given row or column.
* EDIT	E	Modify the content of a given pigeonhole.
FORMAT	F	Format the content of a pigeonhole
* GOTO	G	Position the cursor or box by address on row/column or name.
HELP	H	Get help page from computer.
INSERT	I	Insert a new row or column (empty) in the spreadsheet.
LOCK	L	Protect the content of a pigeonhole from accidental modifications.
MOVE	M	Change the position of a row or column.
NAME	N	Give a symbolic name to pigeonhole
OPTIONS	O	Modify calculations options (interactions, recalculations, etc.)
PRINT	P	Print
* QUIT	Q	Return to OS by abandoning the spreadsheet. Start from scratch.
SORT	S	Sort rows or columns.
* TRANSFER	T	Transfer the content of the spreadsheet on a disc or vice versa.
* VALUE	V	Input constants or formulas into a pigeonhole.
WINDOW	W	Define more windows of the spreadsheet.
XTERNAL	X	Copy from outside into the spreadsheet.

9.3.4　Applications Examples

Function	*Argument*
Introduction of alphanumeric data	*Type "A"*, then type the alphanumeric data at the position of the cursor/box. End with RETURN.

Everything which has been written can be corrected with positioning of the cursor. Thereafter, it is put in the pigeonhole, but the user sees only the part permitted by the chosen dimension of this pigeonhole. (Notice that the two bottom lines of the page are scratchpad.)

Function	*Argument*
Introduction of numeric data or formulas	*Type "V"* Input the numeric data or formula. End with RETURN.
Modification of content of a pigeonhole	*Type "E"* Modify the content of the scratchpad using the functional keys. End with RETURN.
Elimination of a pigeonhole's content	*Type "B"* The computer acknowledges the pigeonhole to be canceled. Example: R1C3 identified Row 1, Column 3. Type RETURN for confirmation.
Definition of symbolic name of a pigeonhole	*Type "N"* Following this, type the name wanted in the position of the cursor/box. Avoid special characters and spaces. End with RETURN.
Addressing of a pigeonhole	*Type "G"* Makes feasible direct addressing through Name "N"; Row/Column identification; Window "W" identification.

After pressing key "R" type:
- number of row

followed by the key S1
- number of column

followed by RETURN.

Depressing the choice "N" recalls the symbolic name (if it is already defined) of the pigeonhole. The choice "W" recalls the number of the page.

Function	*Argument*
Transfer of page content to disc or vice versa	*Type "T"* Diverse choices appear: - Load - Save - Clear - Delete etc.

Type the initial letter of the chosen command:

$$L, \quad S, \quad C, \quad D, \quad . \quad . \quad .$$

"L" *Loads* the page from disc to central memory. The machine demands the name of the file which shall be utilized.

The presentation should be in the form:

$$Y \quad = \quad XXX$$

where:

Y   is the number of the floppy disc (diskette 1.0)
XXX is the File Name.

"S" transfers the page from memory to disc. "C" cancels (resets) the content of the page in memory. "D" cancels the content of the disc.

Function	Argument
Conclusion of a working session, and exit from spreadsheet	Type "Q" The computer asks for confirmation Type "Y" (Yes) in case of confirmation.

It is important to remember to save the content of the memory on a disc prior to exit.

9.3.5 Utilization of Commands

The spreadsheet offers two possibilities for inputting a command of the list at the end of the page.

First, the user can move on the row of commands with the key S1, S2 to reach the wanted command. Subsequently, he should push the RETURN key to execute it. Second, he can type only the initial character of the command.

9.3.6 Calling a Page

Given that the spreadsheet is a generalized program acting like a very high level language, some functions can be used as alternatives. In reality, however, *everyone* has a precise objective. Example:

BLANC	Cancels the contents of a pigeonhole
DELETE	Cancels a row or a column
TRANSFER and CLEAR	Cancel a whole page of the spreadsheet

Very important is the RETURN key. Its usage is necessary following one or more commands to the computer to effect specific functions.

For example, to access the XYZ page we must type:

```
G    ( GOTO )
N    ( NAME )
X ⎫
Y ⎬  XYZ
Z ⎭
R    ( RETURN )
```

With this, the XYZ page appears automatically.

9.4   EDIT COMMAND

Typing the key E or choosing the command EDIT through SPACE BAR and RETURN key, we can modify the content of a pigeonhole.

Through the forementioned procedure, the content of the pigeonhole on which the cursor finds itself is visualized on the working area of the screen. From this moment, the cursor must be positioned on the chosen row and can be moved by using the following keys:

| CRTL | | K | for the movement to the left
| CRTL | | L | for the movement to the right

or by specializing a Functional key. Once positioned, the cursor identifies where we can insert characters simply by typing them, or cancel typed characters through | CRTL | | H | .

Having effected the desired modifications, the key RETURN is used and at this point the visualized row will be reinserted in the pigeonhole of departure and the EDIT operation concluded.

At any moment, the operation can be interrupted by typing | CRTL | | C | .

Though not everybody looks at it as an editing function, the power of spreadsheet programming lies in the ability it offers the end user in correlating pigeonholes among themselves.

Say, for instance, we wish to establish a money conversion procedure between dollars stored in pigeonhole R10C3 and German marks in pigeonhole R13C3. The exchange rate is in R7C6. The exchange algorithm would be:

$$R13C3 \quad = \quad R7C6 \quad * \quad R10C3$$

Had the results of different conversions been stored from Column 3 (C3) to Column 7 (C7), we could write an algorithm for the total in German mark conversions in the form:

$$R18C2 \quad = \quad SUM \quad (R13C3 : R13C7)$$

where R18C2 is the pigeonhole chosen to store the result of this sum. This example shows both the power of a spreadsheet and how it commands the computer to perform its operations.

9.5   FORMAT DEFINITION FOR THE SPREADSHEET

The spreadsheet matrix has:

> 63 columns
> 254 rows

This maximum spreadsheet dimension can be divided into pages. Rows and columns in the same page can be correlated among themselves. They will be updated automatically from the computer in case of modification of contents. For simplification of usage each worksheet can be chosen equal to one page. This means, for example:

> 7 columns of 10 characters each
> 20 rows.

On the screen four further rows are destined to man-machine communication. Such four rows are for utility purposes and are not included in the page contents.

In function of the services offered by a given page, some columns may have a different dimension than 10 characters. For instance, the money column may be 11 characters (making it possible to write up to $1 billion). To the contrary, an account identification column may be of 3 characters.

A noted flexibility of spreadsheet programming is the ability which it offers its user in defining the size of the columns according to *his* requirements.

Furthermore, the user may choose to scroll the columns (and/or rows) rather than page them. Though this adds some flexibility in presenting information which goes beyond the 80 columns of a typical video, paging is a much more elegant approach to page development, retrieval, and presentation.

Paging permits symbolic identification for recall, makes feasible menu selection approaches, and fits with established practices in electronic message systems such as Videotex. It is always advantageous to present the user with a homogeneous operating environment.

9.6 USING INFOPAGES

Let's assume the user has programmed three infopages through spreadsheet.
These are on the ''ARCHIVE'' diskette. Each infopage is integral part of the
spreadsheet to which it belongs.

Let's also assume that each infopage is identified through a symbolic name,
as follows:

Argument	Function
TRETEM	Treasury
PORTEM	Portfolio
DDATEM	Demand Deposit Accounting

These are structural pages (masks, templates). They do not contain data. They
can be copied into other infopages of the spreadsheet (of an equal number of
rows/columns and format) into which data will be inserted.

The auxiliary memory (diskette) can be conceived of two levels of reference:

1. The diskette ARCHIVE containing the File. For this introductory text, a
FILE is defined equal to the whole spreadsheet.

2. Such FILE is defined in terms of infopages (multipage). Each page on the
FILE will be called through its symbolic name.

Let's follow as an example the access to the template TRE (Treasury). To
obtain this page, the user must:

1. Insert the diskettes
2. Switch-on the machine
3. Wait till the cursor lights up (after the light of the diskette turns off)
4. Type:

    ''M'', ''P'' (Multiplan) RETURN

5. Idem, as in point 3, wait until the light of the diskette turns off
6. Follow the command sequence

    ''T'' (Transfer)
    ''L'' (Load)
    1:TRETEM
    RETURN

For practical reasons, it may be more convenient not to recall the template
TRE, but the FILE of the preceding day:

    ''T''
    ''L''

```
1:TREddmm
RETURN
```

As an example on how the contents of an infopage can be stored in memory for subsequent retrieval, say the treasury calculations made on December 8 (1208) must be memorized. The page identification will be:

TRE1208

For the storage of this page on the diskette, the user must apply the command:

```
"T"   (Transfer)
"S"   (Save)
1:TRE1208
RETURN
```

Then wait till the light of the diskette turns off.

File organization on the diskette may follow three main files: Treasury (TRE), Portfolio (POR) and Demand Deposit Accounting (DDA) as identified. The user has the choice of either creating another file with the templates: TRETEM, PORTEM, DDATEM, or using each of these templates as the leading page in the file.

It is quite essential that once this choice has been made, it is consistently followed in the user's work with his personal database. Say that he/she chooses to stick to three files using the templates as the first pages in each file, which can be further subdivided into multipages (Records).

For instance, the Portfolio File may contain templates specifically designed for reporting on domestic stocks, foreign stocks, domestic bonds, and treasury bills. It may also contain a summary template which includes the capability of converting different currency into one reference currency.

For simplicity and consistency in man-machine communication it is advisable to observe a general format of presentation. For instance:

			Transactions as of mm/dd/yy		
Transaction Date	Value Date	Reference	No. of Tran.	Debit	(Credit)
(10 ch)	(10 ch)	(25 ch)	(3 ch)	(11 ch)	(11 ch)

A page can be programmed to accept, say, up to 12 transactions—the 13th transaction going to the next page. Or, the user may decide to augment the number of transactions through scrolling, so that one easily identified infopage in memory is tied to one day of work.

A different way of making this statement is that while programming the computer to do the needed data processing work is tremendously simplified through spreadsheets, the user will be well advised to do the needed organizational work prior to starting his programming. A clean organizational and procedural job is a prerequisite to any successful computer application.

Chapter 10

DISTRIBUTED SYSTEMS IN MANUFACTURING AND MERCHANDIZING

10.1 INTRODUCTION

There are plenty of problems associated with being on the leading edge of technology. Could we master advanced software and new hardware announcements to produce useful applications without delays? Would our staff and users be prepared to handle the resulting systems? Can we make that complex gear work right and keep on working?

Being at the leading edge of technology has, of course, its payoffs: Sophisticated systems attract and retain talented employees. The practice is also cost-effective and keeps the company competitive. The leading edge is much more important because high technology can work for some organizations, but it is not for everyone.

Several firms have exemplified the competitive value of getting into high technology and making it work for them. Boeing took a lead in automating the design, engineering, and production of aircraft, so that it gained in productivity several years ahead of its competitors.

A similar reference can be made about the outstanding needs at the production floor, and in the distribution network. Floor planning is a complex issue. It involves financing and maintaining inventories, keeping track by item number, billing, handling accounts receivable, and so on.

Banks which keep ahead of competition introduce Home Banking (HB), Point of Sales (POS) equipment, automated teller machines (ATM), and electronic tellers (ET) to implement, maintain, and improve round-the-clock service to their customers. Able management allowed several banks to make the investment look smart and attract business.

Distributed systems tie in with office automation because the same workstations serve in both environments. For both references many new applications packages are available for running on PC, that make the implementation fast, economical, and efficient.

Workstations and distributed information systems concepts are fundamental to the success of the new network environment. Whether in the manufacturing industry, in merchandizing, in insurance, or in banking, high technology has two positive effects:

- It provides rich functions to users.

- It allows one to ride the front end of the cost curve, taking advantage of longer write-off periods.

If a company has a growing workload and must continually look for more functionality and more horsepower, economics call for staying on the leading edge. Falling behind in sophisticated equipment can take a toll in any business.

10.2 FUNCTIONAL DESCRIPTION

The best way to characterize the basic structure of a distributed information system is through the principal functions treated at each level: The technical/ factory functions, marketing/sales activities, and headquarters.

Consider an industrial firm with factories and regions with sales offices. The distributed system will encompass:

1. the workstations installed within the factory department and the sales office

2. the computers at the factory and at the sales region

3. the computers at the functional management level, at headquarters

4. the mainframes which are online/offline to the computer networks and for batch processing

Companies involved in the sale and distribution of products perform basically the same business activities, an example being sales analysis and production planning. Together they involve analyzing actual sales from previous years and forecasting for the coming year.

From this information, management plans the production schedule and volume, identifying the stocking point to which specific quantities of goods should be kept.

Scheduling production relies heavily upon the forecasted sales. Each forecast is identifying:

- type of product to be ordered

- quantification of demand for this product

- stocking level to be kept

- economic production lots for product manufacturing.

To answer such processing requirements, the database should contain information on past sales, customers and product activity, including inventory level fluctuation at stocking point (by product), and product pricing.

Another sector of the sales, inventory, production, and distribution cycle involves purchasing raw materials for manufacturing the finished product. For companies dealing in seasonal or semi-seasonal products, this activity is highly dependent on forecasts. The products are manufactured and on the way to the stocking point by the time actual orders are submitted by the sales staff.

Companies dealing with few large customers rely more heavily upon single larger orders placed by these customers, with several deliveries scheduled throughout the year. Manufacturing in this type of business is also dependent upon immediate knowledge of stocking point inventory levels, but production planning follows different criteria.

Product availability will be enhanced when computer systems support these activities adjusting inventory levels with new product status information, and generating online reports on specific product queries.

- ordered goods committed for shipment

- stocking point receipt of goods

- new balances by product

- orders withheld (for products out of stock)

- transfer of stock from one stocking point to another

- follow-up on customer ordering and execution

- shipment of goods to a customer

Any company with a network of factories, sales offices, and warehouses has to get up-to-date dealer inventory information to its branches so that their people can perform inventory inspections. But with batch processing warehouses, factories and offices receive data that is already out-of-date. There is no alternative to an online interactive system, if we wish to strengthen control.

Query into the system should be by product(s), stocking point(s), and customer record. This allows one to learn up-to-the-minute information on inventory levels, products, customers, bills, and accounts receivable.

Streamlining of these operations is so much more necessary as one of the most time consuming and error prone areas of any sales and distribution operation is the order processing procedure. Computer based activities should:

1. accept orders with a minimum amount of input data (enriched by the information in the database)

2. minimize errors by automatically pointing out omissions and incorrect entries to the terminal operator

3. maintain accurate control and status records

A properly designed system should be able to handle several types of orders with the same conversational order processing subsystem. This means processing regular orders for domestic, export, or inter-company customers; holding orders which are not transmitted to the stocking point until all required information is supplied by the customer; handling blanket orders requiring that a number of shipments of the same product be delivered on different dates; including future estimates of orders that are expected at a later time; and holding back orders which cannot be shipped until a later date because of insufficient inventory or other reasons.

Typically, the order processing activity will be initiated with the order entry function. The order entry process should see to it that the system responds to entries, including enrichment and error correction.

Other major subsystems should be order tracking, inventory control, transportation, shipping, inter-company stock transfers, and invoicing. Customer order entry should include the facility of identifying customers and products, allocating the required inventory, and applying the proper charges and remarks.

Order tracking must allow for the instantaneous retrieval of any order to provide for status information, acknowledgements, or changes. The inventory control function must keep track of all inventories by product and size at every warehouse, distribution center, and/or shipping point. Transportation control should account for all physical movement of goods from shipping point to destination.

The shipping entry typically consists of reporting shipments, including such information as mode of transportation, carrier, standard point location codes, commodity information, weight, consolidation considerations, and so on. The invoicing function can be performed at the time of shipments verification or later, provided all the necessary information for invoicing is available in the database.

The whole system should be oriented toward increasing customer service while reducing overhead expense and capital necessary to distribute products. It should be designed to accomplish this objective by monitoring and controlling a customer's order from its beginning until its execution, billing, and payment.

Ideally, such systems should be designed to minimize the variable information required from the user by selecting appropriate data from its storage. For instance, as part of the order entry process it should determine the customer (Bill-To and Ship-To), products that are ordered by the customer on a repetitive basis, primary shipping point, and routing.

The database/data communications system should provide for the consolidation of large groups of business information which may be used by cognizant executives for different purposes. This information may be accessible in different ways, depending on the functionality and purpose of the system components on which it resides.

Objectives evidently include minimizing the number of steps required to perform such functions as order entry, minimizing clerical paper handling, doing away with manual search for information, using the system to edit data for reasonableness and correctness immediately upon entry, capturing the data at the source in a timely fashion, and doing so in sufficient detail to be useful to functional management.

The company must profit from the standardization that this policy will entail. Simultaneously, the sales regions, the sales office, and the factories should be the masters of their applications, while under no condition should distribution in computer resources proliferate the personnel.

Local computer specialists are unnecessary, ineffective, and expensive. A small crack group centrally located can trouble-shoot, whenever the need arises. This will be a reality when management policy is simple and straightforward. Progress takes place from the top down.

"We don't delegate responsibilities for keeping up with what is new. The management searches for new ideas and teaches how to put them in action," said the responsible executive at Georgia Pacific, one of the first large organizations to implement fully distributed systems.

*   *   *

Georgia Pacific is a $2.5 billion company that produces raw wood materials and manufactures building materials, paper products, gypsum, industrial, and consumer products. With headquarters in Portland, Oregon, Georgia Pacific has more than 35,000 employees. The corporation conducts business through:

- 130 sales offices
- 201 manufacturing plants
- 147 building products distribution centers

Prior to the installation of the DIS, the distribution centers utilized a manual system for entering orders from wholesale accounts. Invoices were typed and a copy was mailed to Portland, where orders were keypunched and batched to an IBM mainframe for processing. Following a feasibility study, Georgia Pacific considered

1. installing remote "dumb" or "intelligent" terminals tied to a few central or regional mainframes systems

2. installing accounting machines in each branch for invoicing via mail and key punching or tape reader

3. using remote online/offline small computer systems, with dial-up comunications

The small computers provided the solution: to distribute computing power to the source of the problem—the distribution centers. By capturing order entry on computers at each center and using remote data communications, cash flow was drastically improved and important data made available on a timely basis to the people who neeeded it.

10.3 KEEPING COST-EFFECTIVENESS UNDER PERSPECTIVE

In this and many other cases, feasibility studies established that it is much easier to give the commercial network a dedicated interactive system for the processing of sales orders than it is to add this function to the mainframe. It is less costly to implement an application on a small computer than on a large mainframe. There is less systems overhead, less long distance communicating lines, and the whole solution is much simpler.

Distributed resources see to it that the company gets new applications implemented much more quickly and with fewer systems problems. Management gives the sales force a system responsive to its needs, with good throughput and fast job turnaround. This succeeds in accelerating the whole billing process with very significant financial aftermaths.

One microcomputer-based implementation handles pricing at the sales office level. A price format is displayed by the system. If special rather than standard prices are to apply, the operator keys in the special prices and the system displays subsequent screens where additional pricing or miscellaneous remarks may be entered. The latter can include:

- taxes
- freight charges
- credits
- instructions to the stocking point
- instructions to the carrier

After the last information has been entered by the operator, the system assigns an order number to the data and creates a new order record. The order number is displayed back. The order number and order data are transmitted to the stocking point where another PC automatically prints picking, packing, and bill of lading documents.

Functional executives, who need to check the status of orders which have been entered, can use their PC to display previously entered orders in a variety of ways: Full display of a single regular order or a back order, summary display of a group of orders, outstanding or unshipped orders from a specific stocking point, and so on.

Based on a review of the order(s) it is possible to correct or change an order. For instance, the customer may want to add or delete specific products or alter the quantity ordered.

The terminal operator can change the information in a hold order, if necessary, and release the order for shipment at the same time. The hold order is given a regular order number and stored and processed in the same manner as a regular order.

Through self-correcting features, the system must be able to retransmit to the stocking point an order that was previously entered, but never received because of equipment or line malfunction. This procedure relieves the operator from having to enter the complete order a second time.

Deferred shipments should be scheduled according to requested date, geographic location of customer, and type and quantity of product ordered. The specific products must be subtracted from the stocking point inventory, and the dispatching papers to the delivery point properly prepared.

In addition to supporting the basic business activities being described above, the system should also perform several other duties such as: interfacing with the accounting function; maintaining customer, product, inventory, and order databases; retrieving miscellaneous information for a variety of users; and sending, storing, and displaying messages transmitted among terminals, warehouses, and the central system. It should also handle an automated credit memorandum, regarding line of credit, returned materials, discounts, and so on.

Database maintenance facilities should support the ability to add or modify customer data online. New or changed customer information should be entered, edited for content and completeness, and then written on the master data file. It should as well be possible to edit the data at headquarters by entering special codes and information not available to the order entry offices.

The system should also allow interactive communication between workstations, a WS and the central system, and the like. Each WS should handle both transactions and messages—the latter through mailboxes. Urgent messages must be displayed immediately on the receiving terminal. Even if the terminal operator is performing a procedure when the message arrives, such procedure may be reestablished with no loss of information.

A routine message can be stored in the receiving mailbox until the operator asks to have all incoming messages displayed. Electronic mail is one of the key implementations in office automation and it should be given the weight it is worth.

What has been described is easily within the capabilities of PC based systems. The applications perspectives are simple in their conception and valuable in their implementation. There is a long list of such systems handling order entry and invoicing functions as their main mission.

Striking is the difference between what can be done with the PC and what had been done earlier. In one specific case, under the "old system," orders were shipped, but invoices did not reach the wholesalers for ten days or so because the invoices were prepared at a central location. With the PC, invoices are sent out with the shipments, thus improving cash flow by ten days.

This approach to distributed systems has provided for:

1. improved cash flow by speeding up the processing of orders and billing

2. enhanced control for distribution center managers who now have access to weekly gross margin and sales analysis reports

3. increased speed and efficiency in business areas like (a) order entry, (b) inventory control, and (c) general ledger

4. improved customer service

5. greater flexibility to grow or to change applications

6. increased reliability over the previously centralized solutions

7. ease of use even by novice operators

The benefit from cash flow provides enough financial advantage to cover costs over a brief time period. Though 170 PC were involved in one single order, the cost in purchase price has been about 60% of one year's savings. With software, implementation and telecommunications costs included, the cost was recovered in about a year and a quarter.

Notice, however, that this commerical network not only uses a distributed physical resource, but also it has been projected since the beginning with the aim of handling all applications which originate at the sales office and are treated by the regional marketing management. For this reason, particular care was exercised in designing it independent of the type of workstations to be used and the DIS computer serving them.

The configuration of the installation at the regional office level is kept flexible. Particular attention has been paid to the design criteria necessary to reflect the telecommunications function. This is a direct reflection of the fact that ''any WS to WS'' condition had to be satisfied. The message traffic concerns:

1. the sales orders

2. the credit authorizations

3. the opening of new client accounts

4. the billing

Part of the credit authorizations traffic which exceeds the authorization limits assigned to the regular office, and all new client account messages, will be transmitted during the night to the regional computer. If, for some reason, the regional system is out of order, or the communications lines do not function, the local (sales office) keeps these messages on disc to be transmitted later on.

When the salesman sits down with a client, or even answers a phone call, he can see exactly:

1. what the customer bought over the past time period

2. what the status of the needed inventory is

3. what the margins are, item by item

The manager for his part can access the database and review:

1. each salesman's performance
2. what he is selling or not selling
3. what each customer orders individually, and
4. how profitable each deal is

He can do all this without going through complicated printed reports and manual calculations. With an interactive system he can get the information he needs and when he needs it, right on his own terminal.

Not only have accuracy and timeliness been dramatically improved but, furthermore, the extended use of PC and their topology led to the overall personnel reduction. At headquarters, the data personnel which for over 20 years has been a costly but integral part of the operations, is on its way to extinction.

This is not an uncommon experience among organizations which introduced interactive approaches in their operations. But some companies which eliminated the data entry people and replaced them with professional operators of video units, found that experience unrewarding. Better give videos to the end users and let them do *their* data entry.

10.4   SENSE OF RESPONSIBILITY

The key to the change of spirits and of facts, can be found in the sense of responsibility promoted through personal computing. Here is a basic principle which merits being written in block letters. *The best way to control cost is to make people responsible. The centralized mainframes had made the users alien to a sense of personal responsibility.*

Equally, in the way of disappearance is the heavy computer output. Interactive terminals prove to be far more efficient. Yet, a mere 12 years ago, in the early 1970s, a mainframe based large industrial organization still punched 15 million cards per year. Another, a leading bank, stamped 500 tons of paper each year. Enough paper to snow under all sense of responsibility!

People work best when they have no one to blame but themselves. When they are snowed under paper, efficiency greatly suffers. That's one of the principal issues online interactive solutions aim to correct.

As with all commercial, industrial and financial applications, the network components are basically four: the procedural structure, the hardware resources, the software implementation, and the subscribers (users). Within this environment, all operations are executed in an orderly and timely manner. "I would never try to do all this in such a timely and cost-effective manner, with a central computer," said the responsible director of a firm which has for years implemented a distributed information system.

To keep the operation simple, at the sales office location there is no software handling. Programming is centrally controlled and so is maintenance. There is also, it should be added, a major reduction in the programming effort for the first time after many years of increase in the number of programs being written.

"Simplicity" is the answer to the question "Why?" There is so much you can do with a PC. There is no room for exotic software. The software will take years to write on a mainframe; and only a fraction of that time on a PC. This is fairly well documented by the timetable followed by companies which have worked both ways.

Other benefits come in the way of management control. In the regional office, a larger system will automatically dial-up all the PC and collect information on sales to be used in corporate profit and loss evaluations. Computed data on sales margins is sent back to the branch managers of the distribution points. This distributed system enables local management to find out exactly where it stands at any given moment.

Management reports are prepared at the regions and the branches. The centrally important data is transmitted to headquarters and the equipment there combines text and data for top management. In the case of Georgia Pacific, this distributed environment has given a great boost to marketing analysis and management by exception.

The general case in business and industry today is that many distributors don't know which customers are profitable and which are not. They simply don't have up-to-the-minute sales analyses so that they can spot changing buying patterns or changes in return on investment or turnover ratios on a given item.

Centralized sales analysis can be a complicated business. There is simply no way we can have all the reports we need prepared out of large, unwieldy volumes of data. Monthly reports are prepared on a batch system help, but they are literally out-of-date before they are printed. Information systems distribution helps to focus attention on the area which needs management decision now.

This should not be done in a costly manner, playing with long communications lines. Time and again, the management of clear-eyed companies stressed the need to control costs, to use public lines. Value added networks (VAN) provide a cost-effective communications facility:

1. routing and routing alternatives
2. store and forward
3. logging and journalization, at nodes switches
4. end-to-end message control
5. efficiency in data transmission
6. load smoothing, and
7. reasonable cost

In turn, this permits, at the user side, the development of a peer system with no centralized control. Each node can possess all information which allows it to work independently of the others. It can keep critical data and at the same time interact with the other nodes over the public communications lines.

Data communication takes place as often as required. However, for practical purposes many companies have adopted the procedure that the data link is established no less than once per day and no more than three times per day. The communication between the sales office, the region, the factories, and the head-quarters is daily, but exceptional procedures have been established in case an urgent communication is necessary during the working hours.

With a VAN, the company pays per hour of utilization as opposed to the lump sum of money it costs to run a network of private lines which have to be paid for whether utilized or not.

In one reference, the analysts calculated that given the local data storage and data processing capabilities offered by DIS, data communications volumes will not be increased in a significant way. Some 80% of all data born in the periphery concern the local operations and should stay in storage there.

10.5   AN INDUSTRIAL NETWORK BASE

To reflect the prevailing spirit in terms of system architecture in manufacturing operations, the following outline has been prepared presenting a comprehensive view of a general resource sharing network. Information systems developments closely follow the evolution in manufacturing technology reflected in figure 10.1.

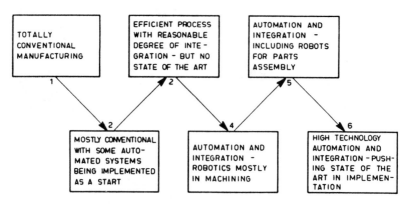

Figure 10.1   Evolution in manufacturing technology

An information network for manufacturing has to provide supervision and control of large production type applications. An aggregate of computer devices will be used to:

- control processes
- synchronize them
- report their status
- provide realtime handling

It should be recording the occurrences of events within each process, reporting them to a supervisory level. To suit factory management objectives, the computers running the manufacturing system should require access to large amounts of data whose location will be determined by frequency of use and ease of access.

Each part of the database should be maintained local to its most frequent use while still being accessible to authorized parties (men, machines, programs) in the network system. A comprehensive set of utilities is necessary to help the user perform routine service functions. One way to identify them is to distinguish six functional groups:

1. system initialization and maintenance
2. resource management and accounting
3. data entry and file review
4. line spooling
5. error logging and reporting, and
6. user communications

The System Initialization and Maintenance utilities will control system start-up and shut-down, monitor system status, set terminal parameters, assign job priorities, and perform general file maintenance functions such as saving and restoring data files.

The Resource Management and Accounting will relate specifically to the user accounting functions, maintaining and controlling status, and allowing the system manager to monitor run time for: user terminals, memory usage, log-ins and log-outs, disc storage allocations, and other peripheral device usage.

The Data Entry and File Review subsystem should permit the end user to design screen formats and assign edit checks to various fields on the screen. The user should have the option of passing each transaction to programs to be processed, or writing the data into a file for immediate access by other system components.

Through the use of the utility-created screen formats, information residing in files can also be reviewed, verified, and if desired changed or deleted with the proper authority.

The Line Spooling routines will permit the user to transfer files from discs to a line printer or other output device. When employing the spooler function, requests are queued and initiated when the output device becomes available or may be scheduled for later initiation. This prevents any contention conflicts which could develop for the system printer.

User programs can then be released to perform other tasks, increasing system performance. The spooler utility can execute other user services. It can print multiple file copies and delete the file after the information has been output.

Error Logging and Reporting features will assure a means of summarizing data or error-related information. The object is to compile and format such data to be output to a terminal or printer. This allows the system manager to obtain a full or partial history of error-related data.

Finally, *User Communications* will handle transactions, summary information, reports, and data files. It will assure communications between "this" system and other computer systems in the network; and generally enhance the efficiency of the data communication function.

As it is so often repeated, the whole idea of interactive computing implies the ability to communicate. This is just as valid among men as it is among machines and procedures. (At the factory level, the applications procedures can be classified into the following broad categories: production planning, production control, inventory management, management statistics, general accounting, analytical (cost) accounting, technical maintenance, personnel and payroll, cash flow, expediting.)

Production planning requires uninterrupted service. The next most crucial applications are inventory management and dispatching. Under this perspective, several approaches have been examined in redistributing data handling—a hitherto centralized solution.

By type of processing—separating communications processing from DP functions, using such physical components as frontends, intelligent terminals, and smart network switches.

In this distribution, the integrated network often becomes the focal point of the architecture. In some respects, this is the easiest type of distribution to achieve because it is relatively transparent to users and does not require applications systems redesign.

By application—isolating cash management, PPC, inventory control, design/engineering, control of machine tools or other distinct application systems, typically using PC or small computers.

The design focus here shifts to the transaction flows either within or among departments. This type of distribution is usually not too difficult to achieve, since many individual applications can be isolated, while also provided with communications facilities to other applications.

By organization unit—designing all applications to match some form of organization structure such as a division, branch office, or plant.

This form of distribution is nearly synonymous with decentralization; and, for this reason—the most difficult type of distribution to accomplish because of associated shifts in organizational responsibilities and the major changes required in information flows.

With each and every solution, fast response time and quick turnaround are at a premium. Although inventories are considered assets, they can also represent a substantial liability. On the average, carrying costs run 25% to 40% of inventory value when we consider interest, handling, storage and other charges. And it is getting more and more difficult to control inventory levels. Yet:

1. An interactive system can not only tell us exactly what we have in inventory right at this minute.

2. It can also help us revise order points to reflect actual usage.

3. It can give local manufacturing management an up-to-the-minute picture of orders and production.

4. It can allocate orders against inventory and work in progress.

5. It can prepare packing lists.

6. It can print out shipping documents and other reports.

Adding to this picture, the PPC requirements permits total control of the production process. In addition, it makes feasible integrated production reports for the entire plant. Production data from each control minicomputer can be fed to the supervisory computer and then consolidated into interactive softcopy-based reports for plant management.

Production planning can be optimized through the use of mathematical models and online terminals exploding the total production (of both components and assemblies) into key systems and single units—matching them against the production resources.

Once the local database is established, product files, component files, and procurement files can be managed through interactive approaches. This will give management a perspective in decision making which has been lacking for so long.

Part Three
USING ADVANCED TECHNOLOGY

Chapter 11

CHALLENGES WITH SEMICONDUCTORS

11.1 INTRODUCTION

A flexible choice of processors and operating environments, loosely coupled architectures with a variety of workstations, but common end user interfaces and data management support, will be among the basic characteristics of future systems. Other features are advanced end user interfaces, user ability to generate applications, increasingly distributed architectures, and object environments able to facilitate distributed processing, provide modularity, and assure ease of maintenance.

At the management level, knowledge processing for decision support systems will be one of the overriding requirements. These are examples of applications areas underlining new demands on chip technology and bringing up questions of choice between off-the-shelf circuits, custom VLSI chips, and new alternatives falling between these two solutions.

The pendulum which a few years back swung away from custom LSI toward the de facto standard of a few very advanced microprocessors, may now be ready to swing back, but with a difference. CAD workstations can expedite chip design giving PC and peripherals manufacturers the capability of devising their own custom VLSI chips in a fraction of the time that it used to take.

As this practice develops, it will induce end systems manufacturers to set up in-house circuit fabrication facilities: the growth of captive IC already exceeds that of mass produced standard chips.

While these changes may characterize the marketplace of the next few years, it is also worth paying attention to the projected milestones in computers and communications in the last 20 years of this century. Starting from the premise that hardware is mindless but can acquire an intellectual focus through software, the following milestones to year 2000 may lead to intelligent machines:

1. Very large scale integration.
2. Universal, inexpensive broadband communications

3. Voice recognition; voice answerback.

4. English language man-machine interactivity.

5. Self-sustaining associative databases.

6. Cryogenic circuitry at biological densities.

7. Holographic memories.

8. HW and SW adaptive to environmental stimuli.

9. Teachable computers.

10. Brain type metalanguage.

11. Computer-based intuition.

12. Brain augmentation through higher level intelligence.

While some of these developments are a decade or more away, it is good to keep an eye on them as they identify the possible direction toward which the semiconductor technology is moving—as well as its aftermaths.

11.2 MICROPROCESSORS

The evolution of microprocessor architecture since 1971 has progressed from the primitive 4004 to the present spectrum of sophisticated microprocessors. They incorporate:

1. Astonishing computational capabilities

2. High-level languages

3. Silicon casting of DBMS and datacomm protocols

4. Technical innovations only recently introduced on larger mainframes.

In the decade since the first 4-bit microprocessor was introduced, these devices have proliferated to such an extent that it is hard to find an application of computers and communications where they are not used or not being considered.

Under present-day technology, a microprocessor is a device that follows a plan—provided it can be specified precisely as a sequence of steps. The state-of-the-art microprocessor can do any information-handling task expressed in an algorithmic (hence, procedural) manner. By the end of this decade, this will probably give way to euristic approaches incorporated first in intelligent machines and then on a chip.

A complete microprocessor design usually involves four components:

1. CPU

2. Memory

3. Interfaces to the peripherals

4. Channels.

The design of the CPU has radically changed in the last decade. The introduction of the 16 BPW (bits per word) and 32 BPW microprocessor machines is altering the applications perspective. The following chronological sequence should be recorded:

1972: 4-bit microprocessors (4040)
1977: 8-bit
1980: 16-bit
1981: 32-bit

Together with the BPW change, microprocessors experienced a significant increase in MIPS. This made the following feasible: larger databases, higher communications bandwidth (buses) and, therefore, new uses.

But the CPU is no more *the central* component of a system. It is just one of the subsystems like the memory, the interfaces, and the channels. Still another subsystem is *software*, which can be looked at as *deferred* hardware design.

As design perspectives evolve, chips are being manufactured to:

• respond to software control

• handle specific user problems

• process and store speech

• cast on silicon a DBMS

• perform signal switching

• respond to datacomm needs of all sorts

• incorporate interface requirements and

• conform to one or more international standards

At the same time, microprocessor-based microcomputers substitute the lower end of the minicomputer range. As they scale up in capabilities, they will be challenged in their former field by nanocomputers.

Novel approaches (along with new structures) will evolve in end usage as nanocomputers step in. The latter will work with interpreters rather than compilers, be dedicated to elementary jobs, and form a hierarchical structure of interrelated components.

Simultaneously, parts of the operating system (OS)—and, eventually, much of it—will be written in firmware, speeding up the hardware/software aggregate and, once again, altering end usage concepts. A system hierarchy may, as well, develop as the lower class of engines eventually merges into the immediately higher one—though a still lower class will fill the gap left by this movement.

This steady progress toward newer technologies is one good reason why we should have a clear idea of the types of microprocessors currently in the market. The other reason is that of choice of implementation. The key single chip microprocessors available are:

1. *The Intel 8086, 80186, 80286.* Designed by Intel, it is second sourced by Mostek in the U.S. and by Siemens in Europe. The 8086 is an improved 16 BPW version of the 8080. Its 95 basic instructions are mainly byte oriented— only a few are 16-bit long. The ALU is 16-bit wide; 24 addressing modes are supported; with clock frequency of 5 MHz, the fastest instruction time is 0.4 ms.

2. *The Motorola 68.000, 68.010, 68.020.* Designed by Motorola, it is second sourced by Rockwell and AMD in the U.S., Hitachi in Japan, and EFCIS in Europe. The external 16-bit bus is multiplexed from the 32-bits inside the engine. A 32-bit ALU has been coded as the user machine. Traps for illegal instructions help reduce software development problems, and allow software compatibility with future hardware improvements. The processor supports 56 basic instructions and 14 addressing modes—with many instructions performing triple functions as an articraft of the assembler.

3. *The Z8000.* Designed by Zilog, it is second sourced by AMD in the U.S., Sharp in Japan, and SGS/ATES in Europe. Unlike the 8080/8086, the Z8000 is not an enhancement of the Z80 and has a different internal structure.

Internal registers allow 32-bit double word operations—but the Z8000 is a 16 BPW engine and therefore are the data and instructions paths. It has 110 basic word oriented instructions, either 16 or 32 bits long. Traps of illegal addresses and instructions help as debugging tools.

4. *The NS 16.000 by National Semiconductors.* Introduced in 1981, this is the latest (and relatively less popular) of the four basic microprocessors. It supports 100 instructions and comes in three versions: 16008, 16016, and 16032. Either of the first two can operate in two distinct modes:

- *Native*, directly compatible to the 16032.

- *Emulation*, providing compatibility to the 8080 at four times the 8080 speed.

With 16 BPW, the 16008 has an 8-bit bus and the 16016 has a 16-bit bus. The 16032 features 32 BPW, and achieves an address range of 32 MBy, through a memory management unit (MMU).

Table 11.1 offers a comparison among the four microprocessor types being considered. Fifteen variables are brought under perspective.

Attention has also to be paid to the newer microprocessor types hitting the market. Developed at IBM's prompting, a largely improved generation of microprocessors comes from Intel. The 80186 is a 16 BPW "board on a chip," compatible to the 8086 and incorporating 55,000 transistors. The 80286 is a virtual memory machine with memory management built into it. It incorporates 130,000 transistors.

TABLE 11.1 A comparison of microprocessors

	8086	68.000	Z8000	16.000
1. Introduction year	1978	1980	1979	1981
2. Basic clock	5 (4–8)	5–8	4 (2.5–4)	10
3. BPW	16	32	16	16
4. Basic instructions	95	61 +	110	100
5. Use of microcode	No	Yes	No	No
6. General purpose registers	14	16	16	8
7. Floating point	No	No	No	No
8. Pins	40	64	48/40	48/40
9. Direct access in MBy	1	16/64	48	16
10. Address size in bits	20	24	16	
11. Interrupt provisions	Yes	Yes	Yes	Yes
12. Stacks	Yes	Yes	Yes	Yes
13. Arrays	No	No	No	Yes
14. Virtual system structure	No	No	No	16032 only
15. Debug mode	No	Yes	No	No

The clock is in the 10 MHz level, but design makes the 80186 twice the 8086 speed; the 80286 four times the 8086 speed. Compatibility with the 8086 processor makes feasible the upgrade of personal computers and other engines while assuring program portability.

11.3 DESIGN AND MANUFACTURING

A main challenge in designing for increasingly higher densities is microminiaturization. Computer-aided design (CAD) has been a tremendous help. The main technical barrier to achieving more functions per circuit is production yield.

Highly complex circuits result in a growing probability of defects, so that until the manufacturing process stabilizes, a greater percentage of the total number of devices must be scrapped. When the cost of scrapping exceeds the cost of saving in subsequent assembly and test operations, the cost per function increases in spite of greater densities.

In terms of cost/effectiveness, the optimal design is a compromise between high assembly costs, which are incurred at lower levels of integration, and higher scrapping costs which come around at high levels of integration.

There are other considerations to take into account. End system manufacturers would like to pack as much circuitry as possible into a given area of silicon, achieving the maximum functional capability on each chip. But when this practice is limited to a single, low-volume application, it is not economical. To help provide a solution, semicustom design approaches, like gate arrays, center on standard chips that can be reconfigured by the end user.

This is a way for users to gain more control over device hardware: In years past ROM were combined with microprocessors to make single-chip microcomputers: logical engines that could be functionally reconfigured by substituting a new ROM with a different program. But this also meant that microprocessors must be surrounded with support chips.

Arrays present a new approach: logic gates are interconnected through minute on-chip metal wiring in accordance with application specifications. This makes chip design a highly involved, multi-step process. The phases include:

- overall functional specification,
- logical operation and checking,
- timing and electrical behavior,
- physical layout of the chip,
- test program development, and
- comparisons between phases to ensure that data integrity is maintained

The observance of these functions during a design cycle extends the time necessary for completion. Chips are built up in layers:

1. Layers from the transistors and other components that comprise the logic gates

2. Other layers isolate or interconnect components.

Patterns for each layer are first etched onto separate templates, the masks. Then, the chip is coated with a light-sensitive material (photoresist) and light is used to project the mask's image onto the chip.

Complex chips require up to twelve masks of an intricate pattern. Interconnection lines may be only a micron wide.

Physically, it has become tedious and time-consuming to come up with the patterns for each mask, though with CAD, lines can be entered into a computer and checked for accuracy.

Because of automated solutions, complete systems can now be put on VLSI chips. But to be successful in design, chip manufacturers must become experts in the fields where their chips will be applied.

These are the main reasons why some basic rethinking of chip design approaches leads to new methodologies and tools. Through them, microprocessor chips that might have taken years to develop can be handled in short time. Automated chip development is not only using software intense approaches and graphic displays, it also increasingly orients itself to these two issues. At the same time, emphasis in the production of memory chips has shifted from LSI to VLSI and competition intensifies in the higher densities.

Whether for microprocessors or for memory chips, design ingenuity is at a premium. The same is valid of manufacturing engineering studies aimed to keep yield up and production costs at a minimum. From initial studies to quality assurance the whole process should be computer aided, with CAD/CAM and robotics being a pivot point.

To meet the design challenges which lie ahead as technology races on, manufacturers develop common efforts. An example is the creation of Microelectronics and Computer Technology Corporation (MCC), a joint profit-making venture of 15 companies, including CDC, DEC, Honeywell, Univac, NCR, National Semiconductor, and Motorola. With a budget of just under $100 million, the company will sponsor and conduct research in computer aided design, integrated circuits, software technology, and advanced computer systems.

MCC will not market products. This task will be up to its individual member companies. It will try to maintain the delicate balance between cooperative research and the competing market aims of its own shareholders. The message filters through that working together, American computer technology can beat Japan Inc.

11.4 THE SILICON COMPILER

At the design cycle, the definition of a circuit starts with a high level description rather than at the gate level. A program is written to manipulate elements able to meet a functional definition.

This new design methodology contrasts to the standard cell approach, which has been described more like a silicon assembler than a compiler. Assemblers perform a word for word translation, while compilers can use macrooperations to inject certain details to the translation process.

The silicon compiler is projected as the tool that could automate VLSI chip design. The object is simple: Instructions are compiled into data that can be used to make silicon chips.

This notion leads toward a dual development:

1. First, a new way of looking at chip design.

2. Second, the dissociation of design from manufacturing, thus creating the concept of *silicon foundries*.

We said that the second pillar of the new approach is the dissociation of the chip manufacturing from chip design leading to the notion of silicon foundries. This is assisted by the fact that compiler companies are cropping up, suggesting that it is only a matter of time before their programs produce the new generation of VLSI chips.

The underlying idea is that if a compiler could automate all the phases of chip design, it would produce pattern generation information elements ready to be fed to an electronbeam machine. This would produce the lithographic masks for physical processing.

The background thinking is rather like software production. Though software and silicon compilers generate different outputs, there are parallels between them. At its most basic level, a chip is no more than a geometric representation which can be expressed in an algorithmic form.

This is why a high level language compiler can produce code for target processes, permitting it to describe and regulate the feed to silicon foundries. MIT, for instance, is working on a project using software routines to describe processes of artificial intelligence, and turning the latter into hardware modules.

Not to be missed in this reference is the great impact this methodology can have on software design. We are in the beginning of an era where computer hardware and software will merge into a single process design-wise, to be described through a very high level language, then manufactured in a silicon foundry.

Current R&D projects indicate that silicon compilation will likely follow a structured, hierarchical approach. Chip design (for hardware and software) will begin with a *floor plan*, and end with the definition of smaller functional cells.

It will be left to the compiler to calculate how much silicon area should be allotted for the different parts of the system; and how data will flow between component parts. The compiler's evaluation of system operation should lead to natural compaction and simplified interconnection.

The hierarchical sense of this design rests on a top-down approach. After the general plan has been worked out, smaller functional cells can be designed as fundamental building blocks for the application at hand. Once optimized, they may be replicated automatically to build up large networks of functional elements and devices.

It is proper, however, to underline that silicon compilers are just now emerging. Unlike the chips we know, which have random and tangled interconnection patterns, hierarchically designed chips emphasize order. Working with regular structures, such as memory arrays and control logic blocks, we are better able to manage complexity. This approach eases the decomposition of a larger project into smaller, manageable tasks, allowing text and complicated shapes to be easily manipulated—through computer aided workstations.

The difference between semiconductor companies in the 1970s has been in chip technology. The difference in the 1980s will be in the industrialization of the design and manufacturing cycles. This will become more pronounced as new types of semicustomized chips (gate arrays and standard cells) are being developed to cut VLSI design costs.

By dividing the design process from the manufacturing cycle, we can mass produce chips that are about 90% complete. Then the end user adds the final wiring to make it a semicustom chip. Secrecy is maintained because the design is kept in the end user's CAD/CAM computer who, incidentally, is also the designer of the last, finishing stage.

11.5 PRICE/PERFORMANCE

Price/performance factors dominate the quest for computer-assisted services. Chip technology shows the way. Motorola and Intel are producing a new chip every two years; microprocessors are available in 1984 that equal the VAX and

DPS6/9X in speed; commodity software development facilities exceed most expectations.

As underlined in the preceding section, software is cast into silicon. At the same time, chips are designed for virtual memory/object orientation in database handling. The same is true of datacomm and graphics.

With a galloping technology and dropping prices, distributed information systems reach the single user. They are also being implemented in heterogeneous environments.

Every one of the PC manufacturers claims different bells and whistles, but the differences in product are not that material. The first important thing is the distribution capability and market access.

The semiconductor market has become the real world version of the movie *Rollerball*: No penalites, no time limit, no substitutions, only endurance. At the same time, competition gets stiffer.

Stiffer competition means the demise of marginal suppliers because there already is far too much contention in this industry. Not only will competition in semiconductors intensify but it will also expand in other areas where microprocessors are leaving their imprint.

An example is the private branch exchanges (PBX). Here AT&T and IBM will be competing with one another but also with a lot of independent, relatively prosperous companies which were not even in business fifteen years ago. Communications facilities will be seeing a major boom; and the same is true of added value offerings from electronic mail and Videotex to voice mail.

Besides cost efficiencies, microprocessors in the executive and clerical environments bring together the benefits of:

- distributed processing
- local databases
- gateway to mainframes/mini
- interconnection through LAN and
- fast response in interaction with a powerful local computer

The best solutions combine the latest in hardware with innovative software. The power behind information systems is in ample supply. Ready-made software, and the price tag for hardware are two issues which really go to the heart of the matter.

The user gets what he pays for, but he also must be sure to pay for no more than what he gets. The role of software must be properly underlined in this connection.

In terms of price/performance, there is *a 20% rule*. The IBM personal computer, for instance, currently provides processing capability at least equivalent to the System/360, Model 30, which was the most popular commercial data processing system of the late 1960s. Yet, personal computers cost less than three thousand dollars, whereas the cost of the IBM Model 30 ran into the hundreds of thousands of dollars fifteen years ago.

Solid state electronics has proven to be one of the greatest technical innovations of mankind. Over the last twenty years the sale of integrated circuits has grown exponentially, and it will be at least another decade before the growth curve of the current means slackens.

Results have been achieved principally through continued reductions in feature size. Computer aided design (CAD) is instrumental, but conventional photolithography is expected to exhaust at about one micron. Beyond that, X-ray lithography looks particularly promising, as does direct electron beam exposure.

Research also stresses better ways to package and interconnect VLSI chips. We need to understand what the limitations are. Future progress rests on our ability to interconnect a hundred thousand transistors on a chip of silicon at a cost which is a tiny fraction of the next best alternative. Chip packaging and interconnection technology are a major economic factor in information systems.

The industry discusses this trend in terms of price/performance which goes all the way from microprocessors to the PC level. Achievements largely rest in the steady evolution of semiconductor technology over a current and a projected 25-year time span (Figure 11.1). Technology provides the means to drive prices down while holding performance constant, or to push performance and functionality up, while maintaining prices.

The choice is to minimize price or to maximize performance. Generally, a yearly rate of 20% has been a reasonable price/performance trend for many products. At 20% annually, the price of today's $50,000 system can be cut down to a tenth, or to $5,000 in only 10 years—while maintaining constant performance.

Following the alternative of minimizing price, the micro vendors have benefitted from the underlying technology curve, creating a new and dynamic class of computer products. However, price/performance severely tests distribution and marketing of ready products.

The salesman trying to earn a living pushing the PC and supermicro may have to sell ten times as many units, possibly by covering ten times as many clients. This is generally an unattractive option to existing marketing forces, and distribution channels shift with technology, price/performance curve.

With personal computers invading the workplace and being well launched in the automation of every desk, the role of the mainframes is changing. The central computer resources are *no more* the processing engines. We should think of mainframes as communications switches not processors, involving:

1. Switching functions for message transmission

2. Database integration and backup

3. Online directory assistance

4. Recovery mechanism(s)

5. Reliability assurance

THE EVOLUTION IN RAM

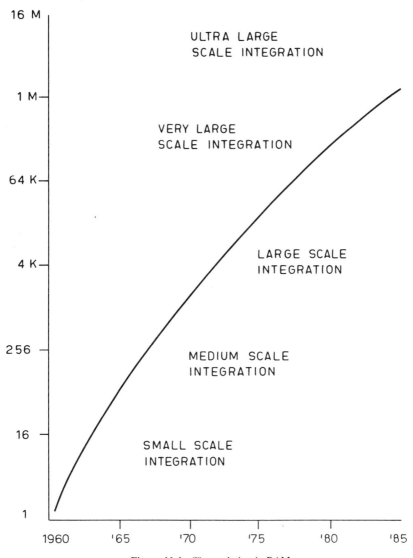

Figure 11.1 The evolution in RAM.

6. Software development facilities

7. Administrative control for the network

Computers and communications technology suggests the wisdom of removing processing from the mainframe and moving it to the microcomputer based workstation. *By doing so, we are making a giant step forward*; toward the effective use of technology.

Is the microprocessor revolution hitting the limits in terms of what can be done through semiconductor technology? While there are reasons to believe that we still have some way to go in this field, the curve will eventually bend. What comes after? The answer most likely lies in Biotechnology and the Biochip.

Chapter 12

BIOTECHNOLOGY AND THE BIOCHIP

12.1 INTRODUCTION

Science and technology have started showing a greater interest in an interweaving of biotechnology and electronics. Biotechnology offers a means of exploiting *bioelectronics*.

As an industry, biotechnology is in its beginnings. In total, it currently offers about 20 products, very few of which have more than a $500,000 market potential. Right now, nobody is producing significant revenues.

The question may then be posed: If biotechnology is an industry that has no earnings and no revenues, why are we interested in it? The answer is linear: Because of its phenomenal potential.

The potential will be realized in three major fields:

1. Health care
2. Agribusiness
3. Electronics

Health care and agribusiness are already well launched in terms of products—though what we currently have available is less than the tip of the iceberg compared to what is expected to come. These two markets are seen expanding in the 1986/87 mainframe.

Bioelectronics will follow in growth curve with about five years distance. This is clearly a top subject for the 1990s and the reason for treating it at this time is to give perspective, to condition our thinking, and to prepare mentally for this colossal evolution.

We can draw a time line. At the far left is time zero. This is the point at which a company asks: "What kind of product do we want to bring to market some time out in the future?" The decision is made, and as we progress along the right of our time line, years pass.

Typically, such time span is riddled with "go/no go" decisions. Management may start a project but then have it fail, so that the idea is dropped. Then a new idea occurs and another project is commenced. That one may proceed for a while and even generate a few offshoot projects.

This process of go/no go is most often internal to a given company. The only time the market finds out is when it enters product trials, and that process can take years. Still, because of the fast developing technology we must learn to look into the future and gauge ultimate product success.

We must also try to evaluate marketing strategy based on experience with current products when they were new (for instance, the personal computer). In these cases, principally we ask: "Are manufacturers planning to use the developing technology in a creative and innovative way?"

If the answer is yes, and the science looks good, this adds up to a higher probability of success. It also means companies are likely to create technological and market barriers to stave off their rivals.

Furthermore, let's keep in mind that the products currently being developed by the biotechnology companies are not necessarily the ones that offer the most commercial promise. They are simply the ones that are the easiest. So we must look at each product in the context of the biological system in which it will eventually function. This entails knowing enough about that biotechnological system to be able to conclude that the product will eventually be used on a broad basis.

12.2 WHAT IS BIOTECHNOLOGY?

The term biotechnology encompasses a wide range of scientific methods. The new genetics works with cells, each of which contains all the information about the growth and development of a given unit (for instance a whole plant).

On this basis, taking agriculture as an example, breeding becomes less time-consuming: Cells can reproduce significantly faster than whole plants. It also requires less earth: whole fields of vegetables can be started in a single laboratory flask.

Still, the foregoing references are only offshoots. The most important issue with *recombinant DNA* is that scientists are able to do more than speed up evolution. They can make hybrids that nature might never create.

This is the core issue behind newest and most dramatic of hybridization techniques, *gene splicing*. First, a gene for a specific trait is identified. Then, using chemical scissors it is inserted into a vehicle, such as a virus, and spliced into a new cell.

Here is where bioelectronics comes into play and opens up vistas of biochip development. Viruses and bacteria are information engines:

- A virus needs 10,000 bits of information to infect others.
- Bacteria require 1,000,000 bits a piece.

- Other microorganisms go to 400 million bits.

- The human organism to function needs a minimum of 5 billion bits.

These 5 billion bits are only a small part of the potential storage in a human being. John V. Nuemann had calculated this storage at the 10^{18} bit level—nearly nine orders of magnitude higher.

Another less known but very important characteristic of information handling in viruses, on which biotechnological research presently concentrates, is the data transfer capability. There are no scientific grounds presently available to explain it. Researchers have, however, established that not only viruses treated by medicaments get immunity, but also they pass this immunity to other unaffected viruses coming in contact with them.

Databasing and datacomm are two of the three pillars on which rests information science. The other is data processing. In this reference, bacteria have already been tailored to produce interferon and to convert waste into alcohol fuels.

Biomolecule immobilization should be one of the key terms in the interaction of biotechnology with electronics. Immobilization can be achieved by chemically or physically associating the biomolecules with artificial support. Two decades of work in this field explain the growing interest in molecular fabrication and manipulation on solid surfaces.

There are evident risks in this work. In a way, genetic engineering is ecological roulette: any mistake will be irretrievable. Some scientists do not object to manipulating the DNA of bacteria to make possible the mass production of insulin for diabetics—but they do worry about other implications (for instance, making changes in the germline, or sex cells, which regulates the transmission of inheritable traits).

Researchers are now working to change the genes in these cells that create inherited maladies. The healthy traits would be passed on to succeeding generations. The abuse of power, once the science is mastered, is the main concern.

Other scientists maintain that genetic research should be encouraged, not met with cries of alarm. Genetic manipulation is seen as man's best hope to eliminate disorders and ameliorate the condition of those affected by diseases. The question though is "Who is going to set the specification?"

While there is no answer yet to many highly relevant ethical questions, enough work has been accomplished to provide background for the technical part. The basic notions are fairly simple.

Expertise in molecular biology is the core discipline of biotechnology. This involves knowhow in biochemistry, protein chemistry, cell biology, microbiology, fermentation, protein purification, pharmacology, immunology, organic chemistry, and electronics process developments at large. It also calls for the application of these capabilities to the cell, the basic biological unit of life.

All living cells contain *deoxyribonucleic acid* (DNA), a complex molecule which encodes the information required for all biochemical processes that occur in the cell. It acts as a high speed memory which is read by the cell's biological machinery to direct its operation. Sections of DNA which manage a specific function are the *genes*. There is a particular gene coding for the production of enzymes, hormones, or other proteins required for the cell's biological processes.

Since the genetic code is universal, the information coded in the DNA of one organism can be translated within another organism. In addition to providing a coded blueprint for the cell's functions, DNA has the capacity to reproduce itself, so that when the cell divides, each and every resulting cell will be a copy of the original and can perform identical biological functions.

Recombinant DNA technology is known as *gene splicing*. It represents techniques providing the foundation for the field of modern biotechnology. These techniques make it possible to isolate genes from higher organisms, including man, and to insert them into microorganisms where they can reproduce and function in the same way as natural bacterial host genes.

This process is referred to as *cloning*. By inserting specific new genetic information into a host cell, scientists are now able to direct the synthesis and production, or expression, or molecules which the host cell would not normally produce.

Evidently, a scientific breakthrough has a market impact only when processes can be industrialized. A definition of this industry is based on the eventual commercialization and development of two fundamental techniques: recombinant DNA, and clonal antibodies.

The fundamental breakthroughs in *rDNA* and monoclonal antibodies that were made in the 1970s have essentially unlocked biotechnology. Until the 1970s, research in those areas was still based on sophisticated observation and some ability to manipulate chemicals.

Today, biotechnology allows the transition from Darwin's natural selection to the manipulation of DNA, the inner core basis of life. The major difference between biotechnology and other emerging growth industries lies in this simple fact.

Like the earlier computer efforts of the 1930s and 1940s, the theories and tools of biotechnology were nurtured in an academic environment. Then the universities began to see their efforts bearing fruit and the products of intensive research have passed into industry. A lot of companies sprang up to exploit the new frontier. Only some will survive.

12.3   A STEADY PROGRESS IN ELECTRONICS

Early computers were designed and implemented with vacuum tubes. In 1948 the transistor was invented. Eleven years later the first integrated semiconductor circuit was built. These are milestones of an electronic evolution, which today aims at 1 megabit chips and has produced microprocessors with 135,000 gates— and which, without doubt, will continue in the future with much higher integration.

This evolution not only has decisive influence on the electronics industry and its products, but also impacts society. The process of industrialization as we have known it during the past 100 years has reached its limits conditioned through volumes, costs, and marketing channels.

Communications is a good example. Though faced with a fast expanding market during the last 60 years, sometime mid-range in this period, communications engineers moved up to a virtual wall made of transmission losses and less than wanted reliability. Only through the use of transistors was it possible to overtake this barrier by increasing the complexity of instruments and systems by several orders of magnitude.

Let's also recall that the first integrated circuits contained only few elemental functions. Yet, their small size, low energy dissipation, compactness, and reliability favored their usage with military equipment and in space programs. Integration also had cost advantages. The more functions are put on a chip, the higher the cost.

How might this development continue? Where are the limits? When will be reached the limit of photolithography with visible light? How far will monochromatic X-rays or electronbeams be used to exposure highly compact structure? The answer to the last question seems to be: 0.01 mm; and evidence suggests that this new frontier will eventually be reached. What's next?

Biomolecules and biosystems seem to be the answer. We should appreciate that our tools of the future are much more complex than avionics, electronics, weapons systems and other processes we master today. At the same time, the possibilities are greater and the development area wider than any we have known until now.

Nucleic acids and proteins form the two most important classes of the biomolecules. DNA stores the information and determines the composition of all biomolecules; proteins perform the living functions.

Nucleic acids and proteins are linear chains. The former consists of four, and the latter of twenty building blocks. Their layout determines both the information in the nucleic acids and the structure of the proteins.

Consider, as an example, proteins consisting of 200 elements. If we form all possible combinations, building only one copy of each, we will fill the whole universe with molecules. A typical protein measures (in every dimension) several nanometers and is an extremely refined miniature engine, working on the border of quantum and classical physics.

This nanosystem has many similarities with solid bodies. Yet, because they are living matter, even simple biological systems can organize, repair, and reproduce themselves. No solid body or chip possesses this property. As a result, biomolecules might be used in electronics, at two levels:

1. Simple polymers serving as elements. Such employment will most likely happen before too long.

2. Whole systems, organized through nucleic acids and bacteria.

Number 2 is much more difficult than number 1 and the field will only then open up when the new physics and chemistry of the molecules will be better understood.

In terms of current research, for instance, there are presently experimental projects on fixing a monolayer of biomolecules on the semiconductor and metal surfaces. A photonelectron transduction has been efficiently performed in the visible light range with monolayer chlorophyll molecules on the model of photosynthesis.

Insight into the complexity of living systems encourages researchers (particularly in Japan) to explore the new frontier of bioelectronics. To their thinking, the molecular linguistic information system could be a reality.

Instead of electrons a messenger molecule moves like hormone and neurotransmitter in the living system. A target receptor recognizes, by decoding, what the messenger means. This process may store supercondensed information and be accessible to the living matter (including the brain and nerves).

An initial stage of this research strategy is to develop a device for electronically stimulated release of messenger molecules and for molecular recognition (for instance, simulating a synapse where the axon of one neuron comes into contact with another neuron).

Quite importantly, in the Japanese experiments the electrically stimulated release of messenger molecule has been demonstrated with conducting polymers such as polyacetylene. The polymer confines a neurotransmitter within the matrix when a specific potential is applied, the neurotransmitter being released by an electric pulse.

Biosensors might be the first practical example of bioelectronic modules. They are being designed to recognize (decode) a specific molecule with resulting generation of electric signal.

• Biomolecules, capable of recognizing a specific molecule, can bind the corresponding molecule to form a complex at a specific site of structure.

• Biomolecules (proteins) are mostly immobilized on the solid support surface which is sensitive and quantitative enough to convert the molecular change into electrical signal. Until today some 2,000 different enzymes from diverse biological sources have been isolated and described. Their average molecular weight is about 100,000.

Let's always recall that the technology of biomolecules is at its very beginning. While we are able to formulate many of the fundamental problems, the solutions which we are seeking are still far away.

We know, for instance, that the arrangement of the building blocks in the primary, linear structure determines structure *and* function of the protein. But we are incapable of predicting, from the arrangement, the structure and function. This makes it impossible to manufacture proteins tailor-made to specific goals.

To scientists this may look as frustrating as the successful manipulation of enzymes, antibodies, and binding proteins. This has been demonstrated in varied manners over ten years. Gene technology also offers great assistance in the refinement of biosensors.

As usual, science is advancing step by step. If we know structure and function, we do not understand the connections one by one. If we suspect that non-linear processes play a substantial role, we cannot measure properly the processes themselves.

On the other hand, imperfect knowledge may eventually prove to be a blessing. Our current thrust to overcome uncertainties has led to the development of different Biosensors:

- the Enzymelectrode

- the Enzymethermistors

- the Enzymetransistors

Their common characteristic is the physical proximity between the enzyme and the signal converter, whereby the latter is the product of the enzyme activities. Their advantages are a direct response and great sensitivity of the sensor based on the very small distance to the signal converter.

Let's not forget that the tendency to bring enzymes and signal converter close together is an emulation of the MOS development. With biochips, proteins are used as electronic building elements. We are transferring, in terms of design, experiences with present-day trends in electronic circuitry.

Oxygen, ion-selective electrodes, enzyme-pH-electrodes, and thermic enzyme sensors (like enzymethermistors) are the most favored signal converters. As most enzyme reactions are exothermic, heat represents the general means of measurement.

Just as important is to underline a rapidly growing interest in instruments which connect semiconductor technology with enzymes or other biological material. The semiconductor elements often used in such studies are FET (field effect transistors). Such building elements have been made sensitive to certain ions or chemical substances, and are known as ISFET or CHEMFET.

In connection with an enzyme, the name ENFET is also used. In analogy to the Enzymelectrode, this building element can be called Enzymetransistor.

It is also correct to make reference to the use of *piezoelectric* crystals, as well as *ellipsometric* approaches. In the latter, no immediate proximity between enzyme and signal converter is necessary.

The principle in ellipsometric solutions is the measuring of polarization angle changes which occur if polarized light is reflected from a surface. The respective change of characteristics (layer thickness) depends on the surface. Hence, a change of the polarization will occur if the surface is covered with a layer of biomolecules.

The addition of biomolecules, which undertake a complex linkage with the layer, leads to a further change. Its extent is measured with the help of a photomultiplier, determining the thickness of the layer.

The idea of using a piezoelectric crystal as microbalance rests on the fact that the vibration frequency of the crystal is reduced when connection takes place on its surface. Research suggests that the same principle might be used in the case of enzymes, which function as an activity layer. This is one of the possibilities currently under study.

12.4   THE NEW MICROPROCESSORS

Based on the considerations we have been going through in the preceding section, we can say that *proteins and DNA are the next great frontier in electronics.* With current integrated semiconductor circuits an increase of complexity of nearly two orders of magnitude, and an increase of the functional speed of about 20 may be possible prior to reaching the physical limits. The way from the myriaprocessor to the supercomputer on a chip for a unit price of $100 seems feasible, based on technological projections.

This is quite impressive but it is not enough. Let's not forget that present-day technology presents speeds and storage capabilities unimaginable ten or fifteen years ago. Yet, for some professions like nuclear engineering and space exploration even the fastest present-day machines are too slow.

At the same time, the procedures for manufacturing and control of structures have not kept pace with product technology. This is even more pronounced in quality control where the checking of big switching complexes is done through algorithms, with the control effort increasing exponentially with greater complexity.

Furthermore, contrary to biological systems, integrated circuits are missing the capability to self-structure and self-repair. Let's not forget that, in principle, integrated circuits are two-dimensional structures.

These are among the background reasons underlining the still tiny but growing interest in *biological microprocessors*, or biochips. The prevailing concept about their structure is that of functionally coupled neurons, and tissue for mechanical and metabolical stabilization.

Granted, the advantage of biological microprocessors, as contrasted to the semiconductor devices, is difficult to estimate because of the rapid development of the latter. But we do know that in the frame of a biological system the different biochips of a central nervous system can be optimized toward specific jobs.

The key issues are design and integration. Though the elementary self-standing biochip will be the first to come into existence, taken out of the context of a nervous system the biochip might only fulfill trivial tasks. We must think in terms of the network.

Each one on its own, the neural elements operate asynchronously, in an analog code. They can show great deviations in their operation and have to be kept in a physiologically adjusted milieu. Biological microprocessors of the complexity of a cerebellum are currently not imaginable (though not necessarily impossible), and a coupling with technical sensors, effectors and other microprocessors will remain a commercially unsolvable problem for many years.

Yet, these references do not exclude all technical usage of biochips. We would be short-sighted if we forget how often technology outpaces our best estimates.

With the Sputnik in orbit, President Eisenhower nominated a blue ribbon committee to study when man could put his feet on the moon. The Delphi method was used and experts suggested that this would happen in the 1990s timeframe, with 1994/95 an object year. But, in spite of the experts, man landed on the moon in 1969, twelve years (not thirty-seven) after the experts had the chance to express their opinion.

Something similar might happen with biochips and we should be aware that a sort of biological microprocessor is being used today in medicine, while molecular clocks are in an up-swing.

In basic medical research the usage of biological microprocessor concepts as neurobiological prosthesis is followed with significant interest. While the nervous system is redundant and plastic, so that often the functions of damaged structures can be taken over by other regions, a replacement of vital substructures through implantation of a biological microprocessor is exciting news.

Technically, such an approach is facilitated through the absence of normal blood/brain reactions. The central unsolved problem is the functional connection of the implanted biological microprocessor neurons with the nervous system.

Cut-off nerve channels of central neurons do not regenerate in an adult. At the high connectivity of the central nervous system it seems necessary to create an estimated 1,000 input/output connections per neuron. Still, initial successes in the implantation of embryonal biological microprocessors suggest the ability to create some functional outputs.

The interfacing of biochips also leads to basic questions of neurogenesis and plasticity, an important area of neurobiology which is developing quite rapidly. Another area of considerable interest gravitates around the knowledge of a molecular clock.

The fact that DNA can serve as a *molecular clock* marks a major evolutionary turning point. Using this method, scientists are rewriting the history of many species, including man, proposing a radical new basis for the system of scientific classification devised by Linneaus over 200 years ago.

Molecular studies show, for example, that DNA from human beings and chimps is about 98% identical. Such degree of similarity usually occurs under this system only among animals of the same genus. Yet men and chimps are divided not only into separate genei but even into separate families: Hominidae and Pongidae.

The idea of a molecular clock is relatively new, but that of charting biochemical kinships among species was originated at the turn of the century by George Nuttall, a British doctor. Nuttall injected rabbits with human blood, causing their immune system to make antibodies, just as if the foreign blood were an invading bacterium or virus.

Proteins, modern science suggests, may function as a molecular clock. Such a clock, according to the prevailing hypothesis, is set up in motion when two animals diverge. New biochemical techniques prompt researchers to apply this principle to the study of not only many animals including humans, chimps, gorillas, apes, monkeys, but also birds, mammals, amphibians, fish, and even bacteria.

Studies show that though proteins often behave in a clocklike fashion, different proteins evolve at different rates. One protein represents only one gene, and one gene only about a very small fraction (two millionth) of all the information in the DNA.

For this reason, DNA hybridization aims to study DNA in its full span. This approach is based on the fact that the double stranded DNA helix can be broken into its complementary single strands by heat.

The procedure rests in combining samples of DNA from two species, boiling the mixture to break the double strands apart, and letting it cool so that the strands will recombine into pairs.

- Each strand tries to find its exact complementary partner.

- Some strands pair with partners from the other species that are not quite a perfect match.

- These form hybrid double helixes that are not as tightly bound together.

Researchers contend that the DNA clock must be calibrated with well-established paleontological or geological reference points. Other researchers, supporting the molecular method, argue that this is much better than the study of fossils for inferences on the family tree of the species.

The thesis of the latter school is simple: "We know our molecules had ancestors." The counter-argument suggests that the molecular clock gives only part of the picture and that we have to combine the research on the ancestors (DNA clock) with that of the descendants (paleontology).

While the interest in molecular clocks as research agents is very real, we should not forget that clock action has been, since the beginning, a basic building block of computing machinery—particularly for synchronous equipment.

Thus to the databasing, datacomm, and data processing capabilities we can also add the indispensable clock action.

With this range of facilities in mind, we will be well advised to forecast in bioelectronics an evolution similar to the one experienced by the other implementation fields of recombinant DNA (for instance, disease treatment). In some companies operating in this field, product strategy focuses on short, intermediate, and long-term goals designed to assure the introduction of a steady stream of new products throughout the next decade.

The First Generation has been diagnostic for Infectious Diseases and Cancer.

Laboratories are developing products for rapid diagnosis of a broad range of human infections. This includes respiratory diseases, such as pneumonia; debilitating infections that occur in burn victims and hospitalized patients with long-term illnesses, like staph; and other diseases such as herpes.

The Second Generation addresses itself to Automated Blood Cell Typing.

Scientists are working on products for the automated typing of human blood cells. These will be used in clinics and blood banks to determine compatibility for blood transfusions and organ transplants, susceptibility of individuals to certain diseases, and paternity in legal cases involving child support and/or custody.

The Third Generation will be Treatment of Infectious Diseases and Cancer.

This effort includes monoclonal antibodies of human origin for the treatment of problematic bacterial infections that commonly occur in burn victims and hospitalized patients with long-term illnesses. In parallel, scientists attempt to develop human antibodies that are therapeutic for the most common forms of cancer.

In a similar manner, biochips will most likely start being used as advanced components within existing systems, slowly developing into elementary products. Next will come more complete offering serving specific markets, small system products and packages of a generalized nature—prior to reaching big system dimensions enriched (most likely) with artificial intelligence.

12.5 BIOTECHNOLOGY AND HEALTH CARE

Significant capabilities in recombinant DNA technology are the product of gene synthesis, protein microsequencing, cloning, and the development of gene clusters responsible for controlling series of chemical reactions. With a focus on health care, four areas are outstanding:

1. Therapeutics, including antivirals and growth regulators.

2. Diagnostics with immunoassays and DNA probes being the key references.

3. Preventive health care where vaccines dominate.

4. Specialty chemicals, particularly substances produced by hydrocarbon oxidation.

Within each product area are alternative approaches to developments, falling into main classes starting with the production of natural gene products by traditional recombinant DNA methods. In diagnostics, this technology refers to the development of assays based on conventional monoclonal antibodies, and in specialty chemicals it applies to the control of single chemical reactions by single genes.

Reference to health care products is a necessary supplement to the discussion of biochips for two reasons: The first reason is to supply evidence on what has been achieved through biotechnology (agribusiness is the other example). Second, because the health care field is the most likely to see first the implementation of biological microprocessors.

Synthetic insulin is the first genetically-engineered product to be cleared by the FDA for human use in the United States, but this is only one of the possible examples.

As viral infections represent one of the most serious and frequent forms of contagious disease, biotechnology companies are developing products for both their treatment and prevention.

Interferons are natural proteins which appear to inhibit viral replication as well as abnormal cell growth. They have been used experimentally in the treatment and prevention of selected viral diseases, as well as in the treatment of cancer. The results of preliminary clinical studies are encouraging.

Natural leukocyte interferons are produced from white blood cells. They comprise a family of related proteins with similar properties, but structural differences that affect the biological activity of these interferons.

Significant work is also invested in developing products for preventive antiviral medicine, specifically subunit vaccines. Viral vaccines currently used in health care are generally prepared by propagation of viruses in cell cultures. The viruses are then administered as live, weakened viruses or as killed viruses.

Problems often associated with conventional vaccines include the risk of weakened viruses reverting to a virulent state. The biotechnological approach involves the use of virus subunits as vaccines. These incorporate external structural elements of viruses, isolating or constructing the gene for the structural protein. Recombinant DNA technology is then used to engineer microorganisms which will produce the subunit protein molecules by fermentation without containing infectious viral DNA.

Along a similar line of reasoning, biotechnology companies are working both on diagnostic tools and therapeutic agents. The use of monoclonal antibodies in diagnostic tests can improve the accuracy and reproducibility of existing tests, as well as accelerate the development of diagnostics not currently available with existing technology.

Diagnostic products for respiratory diseases, such as pneumonia, strep throat, and several forms of the common cold, are being developed. The sensitivity of any diagnostic procedure depends on the binding ability and selectivity of agents designed to react with the target molecules. In contrast to a monoclonal antibody which binds to a segment of the target protein, a hybridization probe is a segment of DNA which binds to a portion of the gene or the messenger RNA producing the target protein.

Diagnostics through the use of biotechnology proceed in the following way. A patient's sample consists of DNA comprised of two complementary strands or chains, each made up of a series of nucleotides.

The strands twist together to form a double helix, the base of most living organisms. As stated, when the two strands of DNA are separated (under certain heat or chemical conditions) and immobilized, two polynucleotide chains result, each with a stretch of unpaired bases.

A positive test result occurs if a double helix is formed when the probing agent is added to the patient's sample. To form a tagged double helix, a patient's sample must contain the identical genetic material found in the probing DNA.

In both diagnostic and therapeutic facilities, a focal point is cancer. Significant research is directed on five major issues: leukemia, cancers of the breast, lung, prostate, and colon. These forms represent 56% of all cancers that occur in the United States.

Biotechnological research also includes studies of cancer viruses, oncogenes (the genetic factors responsible for the cause of cancer), and cell growth factors which influence the behavior of cancer cells. Recombinant DNA technology has also resulted in the generation of hepatitis B antigens from which may potentially be produced inexpensive, safe vaccines.

Another broad area of interest is the production of chemicals by fermentation. It is achieved at approximately room temperature and atmospheric pressure and has environmental advantages over conventional chemical manufacturing processes which often require high temperatures, high pressures, and toxic catalysts. Fermentation produces less air pollution, less water pollution, and virtually no materials requiring hazardous waste disposal, providing many byproducts in addition to the target product which must then be separated.

In the speciality chemical field, initial efforts have focused on the production of enzymes which are the products of single genes. Broader application to compounds that are products of a series of chemical reactions requires multiple genes which function together, is the subject of further research.

Throughout these references, the impressive issue is our ability to manipulate the code of life. This is a fundamental discovery, leading to the use of new, man-created resources, and to significant ethical and evolutionary risks.

12.6 IMPACT IN THE AGRIBUSINESS

Of all fields of applied biotechnology, agribusiness has so far attracted most work. Throughout the United States laboratories are splicing genes, cloning cells, and mating the previously unmateable, with the intent of creating more nutritious, better-tasting, and less-expensive food—as well as to prepare growth hormones and protective vaccines for animals.

In a world where the current population of four billion is expected to double in less than twenty years, the agricultural applications of biotechnology might be even more crucial than medical uses. The agribusiness potential could be by order of magnitude larger than the medical potential, with a market of $50 billion by the year 2000.

Test-tube farming is still in its infancy. There are roadblocks: Not only must genes be isolated and crosses made, but the cells must be grown into living, seed-producing plants that will pass on selected traits to future generations. However, other applications start blossoming.

Scientists have combined molecular biology, plant tissue culture, and conventional plant breeding in the development of novel field corn hybrids. The new proprietary plants promise economic advantage to the seedsman and farmer.

Biotechnology is mastering the regeneration of entire plants from small clusters of cells in culture medium. Large populations of most of the agronomically important field corn varieties can now be grown, genetically modified and screened for novel traits year-around in tissue culture.

This compresses the time required to develop new crop varieties compared to conventional approaches. At the same time, biotechnology addresses short-range opportunities in the animal healthcare marketplace.

• Vaccines, antibodies, and animal hormones are in demand.

• Agriculture has become a technology business, where the market demand for new products is great.

• Genetic engineering can provide improved products and at the same time lower production costs.

This is enhanced by the fact that the regulatory approval cycles for many agricultural products are short, the capital needs for manufacturing are often relatively small, and the market is generally accessible to new entrants.

Regenerated plants are self-pollinated in indoor growth rooms and the resulting seed channelled into plant breeding programs. At least one company established its first permanent breeding station in early 1983 and is planning to establish other stations in the Corn Belt.

The plant breeding staff not only evaluates the genetic variation of regenerated plants, but also selects and monitors new materials to be returned to the laboratories. This research can be instrumental in developing several new varieties of field corn—including hybrids with enhanced protein quality and resistance to new high-performance herbicides. It can also raise the quality and value of a farmer's crop and reduce his production costs.

Different approaches are followed in reaching this goal, but for all their diversity the experimental techniques share one goal: The strategic manipulation of plant genetic material, helping to:

1. Increase crop yields through the incorporation of resistance to drought, plant pests, and the like.

2. Raise crop value by means of improved nutrition.

3. Lower production cost by eliminating or radically reducing soil treatment.

4. Provide early diagnostics and therapeutics where disease persists.

Multimillion dollar losses are incurred yearly by plant and animal diseases like potato viroid and scours. Early detection of the infectious microorganisms causing these disorders is critical to successful treatment as well as to selecting effective control measures to prevent infection of healthy plants and animals.

Current biotechnology helps in detecting a number of infectious disorders, including plant viroid, virus and bacterial diseases as well as farm animal illnesses. One of the agricultural tests being developed is a diagnostic procedure for potato viroid, an infectious agent and serious crop disease. To help assure healthy crops, seed potatoes are screened for viroid at several stages of production.

Biotechnology companies have also directed their efforts toward a group of potential products targeted at specific livestock markets. These include: diarrheal diseases of livestock, respiratory diseases of cattle, viral diseases of swine, and others.

As stated in the introduction, the range of products currently marketable is thin. The period between now and 1986/87 will be characterized primarily by laboratory research. Yet, what we are seeing is the beginning of a new epoch in agriculture.

* * *

Today, applications of biotechnology by and large focus on agriculture and medicine. It is only recently that serious interest has developed in the scientific community on the possibility of using carbon-based materials for computing.

Research on biochips has been a fringe area, but that's beginning to change. As an example, Carnegie-Mellon University has formed an interdisciplinary Center for Molecular Electronics, expecting to have a staff of 15 faculty members and 35 graduate researchers. At the same time, the Strategic Defense Initiative of the Department of Defense looks at computer circuits made of organic molecules instead of silicon—while there are increasing rumors that IBM is doing considerable research in biochips.

Chapter 13

INTERFACING THE MAN TO THE MACHINE

13.1 INTRODUCTION

Computer power will be relatively limited unless it can be easily accessed by the end user. The most user-friendly way of communicating with a machine is voice. But voice media are still at a development stage.

For this decade, the best answer we can provide to the input problem centers on the choice of cursor control devices. Units that fall into this category include the mouse, light pen, track balls, digitizing tablet, joy stick, and cursor control keys. Each has its advantages and disadvantages. For instance, the wire (tail) is a nuisance with the mouse.

The problem of the connecting wire is common to several devices, including keyboards. One of the growing requirements is to make them wireless—something which is already being done in the case of keyboards and joysticks. One interesting problem this raises is the likelihood that small portable devices would naturally go into pockets and become implicitly non-sharable, hence, personalized.

Another problem with the devices we are considering is parallax. The user might not always get what he thinks he is pointing at, because of the various angles involved (light pen to screen, electron beam to screen, and so on). The use of a flat display in a nearly horizontal position should eliminate this problem.

The primary advantages of joysticks are the reduced hand motions and the natural type of the interface. However, for the time being joysticks are not accepted as part of the business world.

The concept of joysticks does, however, raise the suggestion of replacing the four cursor control keys to slide (or respond to pressure) in at least four different directions.

Additional issues are connected to specific devices. Digitizing tablets raise questions of scaling, resolution (of the pointing device), and relationship with windowing.

Scaling is necessary with potential hardware assist. Resolution of the device need not exceed that of the screen.. Windows are a system problem.

Still another disadvantage is that while all of the devices under discussion tend to be cursor control (pointing devices), with the exception of the light pen, they cannot be used to point directly at an object on the screen.

The user must have varying degrees of skill to position the cursor. Pointing a finger may be much more efficient and require no skill. This is also the advantage with voice commands.

The use of a touch sensing screen can help improve both the speed and ease of use of the man/machine interface. One of the simplest problems to contend with is the need to inform the user of the capabilities of the system and allow that person to select a desired function simply and quickly.

The use of menus, both soft and hard function keys, keyboard overlays, etc. have all been used to satisfy the forementioned. Touching a displayed picture (an icon) of an office device such as a file cabinet can be used to indicate a desire to file.. Menus, function keys, etc., can be employed in a similar manner.

One problem with touch panels currently is lack of resolution. Furthermore, while it is possible to display large amounts of options, the view of those options will be partially masked by the user's hand during the selection process.

This is important in that one would like to give the user immediate feedback regarding the function selected. Such feedback is often done by increasing intensity or blinking the object selected. Another drawback is that the screen will get covered with fingerprints.

One possibility under study is a touch panel separate from the screen to be placed horizontally or at some other more convenient angle on the user's desk. Both options and feedback could still be displayed on the screen. However, this is much the same as the use of a digitizing or graphics pad.

Membrane keyboards are similar in nature to touch panels and are already available and they come in a variety of forms. Their use in the fast food industry is an indication of their simplicity and the speed of interface they provide.

System-wise, a major advantage can be obtained by using closely related technologies both for input and for output. Light pens, mice, digitizing pads, cursor control keys, joy sticks, track balls, and different types of panels should be examined in this general context.

One advantage that most of these devices have over the touch sensing screen is that one or more function keys can be built into the device itself. These are used to modify or shape the function being selected.

For example, pointing at a function via a cursor controlled mouse might cause that function to light up on the display screen. Pressing one of the attached function keys might cause the execution of the selected function. Two or three keys seem sufficient for most functions projected out of present experience.

13.2 FUNCTIONAL COMMANDS, CURSOR, MOUSE, AND JOYSTICK

On the videoscreen there will be a *Cursor*. Usually, this is a blinking block or underlined character. The cursor is nearly always on the screen when the machine is turned on. It marks where the next text entry or command will appear.

Different equipment manufacturers use different symbols for the cursor, with a full rectangle or arrow being the most favored signs. Other de facto standard symbols usually look like the following:

Page Format	#—Page Format
Special Print Format	˄
Heading/Footing	#—Heading or Footing
Carriage Return	<
Indented Tab	>

Prompting solutions, as those often used with word processing, present a set of commands fixed through functional keys. For instance, an *Execute* command tells the hardware/software aggregate to "go ahead." A *Cancel* function allows us to "escape" from any entry; it cancels the entry when we have made a mistake and is used when we want to go to another mode.

Similarly, other functional keys will call frames for Page Layout, Tab Rule, Overtype or Insert, and so on depending on equipment functionality. For instance, the availability of an "Insert a character" key permits one to insert a character at the cursor location when editing panels, and while in overtype; an "Erase to end of field" function will delete all the data from the cursor location to the end of the field.

Functional commands are very helpful and can significantly ease man-machine communication. For example:

- *Append*—Appends one document to another

- *Attach*—Attaches an index file

- *Center*—Centers a line while in Insert Mode

- *Columns*—Enters the column heading procedure

- *Copy*—Copies text to the disc for later recall at any point in a document

- *Delete*—Deletes unwanted text

- *Edit Marking*—Marks the location of any edits to the document

- *Erase*—Erases text from the cursor to the end of the document

- *Exchange*—Exchanges text with new text

- *Finish*—Exit the document and store it to the disc

- *Footnote*—Enters a footnote

- *Get Block*—Retrieves a block of text that we have moved or copied, and enters it into the document at the cursor location

- *Hard Space*—Temporarily disables a word wrap

- *Header/Footer*—Enters a header or footer at the cursor location

- *Hide*—Hides or reveals the format tokens on screen

- *Hyphen*—Starts a hyphenation pass
- *Insert Text*—Allows the user to enter text into the document
- *Jump to Page*—Takes the user to the top of any page
- *Move*—Moves a block of text off the screen and holds it until we recall it
- *Next*—Scrolls to the next screen
- *Overtype*—Allows the user to type over existing words and text
- *Page Break*—Enters a hard page break
- *Page Layout*—Enters new margin settings
- *Phrase Insert*—Inserts a phrase from the phrase library
- *Print Style*—Enters a new print style
- *Quick Search*—Searches for any material contained in brackets
- *Replace*—Replaces words, phrases or symbols throughout a document
- *Tab*—Calls the tab rule to the screen
- *Widow*—Activates widow control

The most usual commands for cursor movement are:

- *Up Arrow*—Moves the cursor up. It may also scroll the screen backward if it is in the upper left position on the screen.
- *Down Arrow*—Moves the cursor down. It will also cause a single line scroll when it is at the bottom of the screen.
- *Fore Arrow*—Moves the cursor forward. It will allow the cursor to move out of text.
- *Back Arrow*—Moves the cursor backward. It does not allow the cursor to move out of text.
- *Top*—Moves cursor to top position
- *Bottom*—Moves cursor to the bottom of the screen

Though the cursor is a mature feature having appeared with the early video-screen (softcopy capability), its handling has classically involved keyboard manipulation. This is not a user-friendly solution.

Until voice input/output media are effectively developed and implemented for daily use at workstations, video terminals and simple programming languages help bridge the gap, with devices like light pens, graphic tablets, track balls, joysticks, and paddles.

A track ball is popularly known as the *mouse*. Moving the mouse causes the cursor on the screen to move directly to the desired location. This is a simpler approach than pressing keys on a keyboard.

The use of a mouse as a pointing device is no cure-all solution. The mouse is useful for spreadsheets, and for anything relating to menus and windows.

It is not recommended for applications in heavy-duty word processing. Here, solutions must be provided able to confer enough technical benefits upon their users.

Examples of these solutions are as follows: Strong communications capabilities, facilities for building help screens, menus (with windows being a valid reference), and portability. In other terms: features make the tool an excellent base on which to build the applications perspectives.

However, a little more sophisticated solution may include on the same casing of the track ball up to three special keys:

1. One to move the cursor

2. A second to do scrolling

3. A third to open windows on the softcopy

(Star of Xerox has three keys; Visi-On uses the No. 1 and No. 2 keys; Lisa has only one key.)

Because it is more user friendly than other media, the track ball starts being a widely used device with workstations. It combines features of a paddle and of a joystick, adding some of its own. Its major advantage is that a cursor can be moved rapidly or slowly to any point on a screen with precision.

Part of the ball extends through a round hole in the top of the module, allowing it to be turned or spun in any direction with the fingers or palm of one hand. The ball rests on three points: two steel axles and a ball bearing.

At one end of each axle there is a disc, and in each disc there is a ring of holes. Two small lights shine through the holes in each disc, and the beams cast by all four lights are picked up by light sensors. As the discs rotate the light beams are translated into pulses of light counted by the sensors.

This data is used to position the cursor. Signals from the horizontal axle help determine the vertical movement of the cursor. Those from the vertical axle control the horizontal motion.

The faster each disc spins, the faster the light through it pulses. From such rate of pulsation, the computer can tell how fast to move the cursor.

By noting which sensor acts, the computer determines the direction of rotation and can move the cursor accordingly. Data determining position with respect to a horizontal and a vertical axis is sufficient to define any point on the screen selected by the end user.

Let's recapitulate: Track ball functionality can be enhanced through functional buttons: One, two, or three buttons, are the usual case—depending on the sophistication of the hardware/software aggregate. They add to the performance of the mouse by commanding specific tasks on the screen.

A joystick employs two basic components: a stick and a trigger button. Through the stick, a player can move a cursor in different directions: up/down, left/right, or along either diagonal.

A joystick, however, has its limitations. For instance, it cannot control the position of an object as precisely as a paddle can. There are two main varieties of joysticks:

1. Switch-type sticks, which can move a cursor in eight discrete directions

2. Analog sticks, which offer finer directional gradations

A typical switch-type joystick has four contacts in its base. When the user pushes the stick in a given direction, the part that extends into the base closes a contact; two contacts, if the motion is diagonal.

The software generates a corresponding cursor movement on the screen as long as the contacts are held closed. By contrast, an analog joystick has two potentiometers in its base, mounted at right angles to each other.

One potentiometer turns when the stick is moved forward and back. The other turns when it is moved left and right. The computer then reads the analog signals from the potentiometers and converts them into digital data as it does in the case of paddle potentiometers.

13.3 IMPLEMENTING SOFTCOPY WINDOWS

One objective of integrated software is to smooth the differences between various personal computers. Another, to offer a consistent user interface between all products which may be simultaneously used.

Through an electronic cut-and-paste feature, data may be lifted from one application in one window and laid down neatly into another application in another window. This feature is particularly useful in the integration of spreadsheet calculation into graphs or standard word processing documents.

Windows are independent format handling devices which respond to standard input/output calls, as well as commands to manipulate attributes such as:

- location

- size

- font-usage

- exposure

- keyboard status

Typically, a window is associated with the raster-scan device upon which it is displayed. Windows may or may not have a one-to-one relationship with processes. Every window has an independent multi-font map.

A window clings to a process-group to which a signal is sent whenever the window changes size or location, becomes exposed or covered, gains or loses current keyboard window status, and so on.

Only the user has control over the windows in the system; but a window can be manipulated by a process in its processgroup or by a process with write permission on the window. All windows, upon creation, are entered in the file system as character-special devices; in the directory they have a filename specified by their user-supplied label.

Windows are dynamic in that their location, size, exposure, and font-map can be modified at any time under user/program control. When a window is created, it can be used in a variety of ways by the process that created it. When a window is covered, output to it can be saved in a buffer and displayed when the window becomes exposed again. This way a process will not be halted because it wants to output to a covered window.

Windows lead to the implementation of a simulated virtual memory environment allowing combinations of applications to run concurrently which would never fit into ordinary personal computers.

Other features can also enhance the PC capabilities. For instance, bit mapped graphics features permit vertical and horizontal smooth scrolling of initial documents and graphics images which facilitates viewing items which are larger than the videoscreen size being employed.

User creation and control of windows is accomplished through a set of calls. These allow a user-process to:

- make and initialize a new window
- draw or erase a window
- insert a selected window
- obtain the current state of a window
- nodify the current state of a selected window
- select and manipulate the fonts utilized by a selected window
- read the state of the mouse device in a selected window
- obtain the current state of the display to which a given window belongs

- switch keyboard input to a selected window

Through the call mechanism, windows can be manipulated dynamically under user-program control.

Windows may occupy independent and changeable rectangles of videoscreen surface and may overlap each other. From the system's viewpoint, windows belong to a flat space where the status of any one of them is essentially independent of the status of any other window.

Should a user program wish to impose a hierarchy on a set of windows, by having, for example, the size, location, or exposure of some windows tied together, the call mechanism provides the ability to do so.

For each video display, there is always a current keyboard window to which keyboard input from the user is sent. This current keyboard window can be selected under program control through a call or a special key combination.

For each window, the system will keep track of what other windows are currently covering it. In addition, if a window's save-image user-flag is enabled, the system may maintain an ASCII image of the window's current contents.

Hence, the system can automatically-expose and refresh the image of a window that becomes uncovered as a result of the erasure of all covering windows. For font management, each window in the system may have an independent and changeable multi-font map which associates slot numbers with loaded fonts.

Windows help users easily move between spreadsheet, word processing, graphics, and other packages.

The use of windows in WP/DP applications will succeed as a standard or near-standard for three reasons. First, it is technically the right idea. It permits development of programs without having to write logic. Second, it eases considerably the man-machine interfaces. Third, it does away with keyboard functions which, for executives, are difficult and frustrating.

Typically, in a window system, the mouse is always attached to the current operating position. The mouse driver does allow the user to directly open the device and thereby obtain access through the mouse.

If a user-program wants to paint a window directly, writing characters or vectors into the raster bit-map, a set of library window graphics' routines should be available for this purpose. To assist the window system user in creating new windows or manipulating already existing ones, a window-manager program must be written, making use of the mouse to select functions from a set of displayed menus. Among functions to be supported by the window manager we distinguish:

- the loading, setting, or clearing of fonts from windows
- selection of the current keyboard window
- exposure or erasure of windows
- modification of windows
- creation of new windows

Routines that implement the window device abstraction should preferably be hardware-independent and portable across machine implementations. The interface between these routines and the window device code must be fairly simple and straightforward.

13.4   TOUCH SENSING SCREENS

Touch sensing videoscreens provide a simple and relatively effective way for communicating with computers. Rather than pushing buttons, switches, or keyboards, the user points to (touches) a computer-generated display on a terminal screen. Such screen must be equipped with a touch sensing device.

Through this approach, the computer determines the spot on the screen which has been touched. It then takes the action for which it has been programmed.

Typically, such terminals use a pattern of closely spaced, horizontal and vertical infrared beams which pass very close to the face of the display (Figure 13.1). As a finger touches the screen, the beams corresponding to this spot are interrupted. This permits the computer to identify the touched spot's exact coordinates.

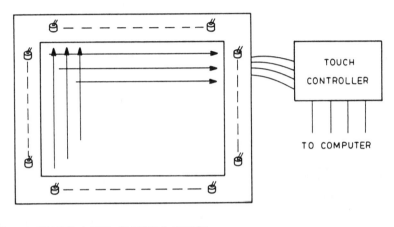

INFRARED LIGHT–EMITTING DIODES
AND PHOTODETECTORS

Figure 13.1 Optical array of infrared light-emitting diodes and photodetectors.

In different terms, the key to touch sensing is the matrix of infrared photodiode transmitters and photodetector arrays that frame the video terminal screen. Sensing the interruption of the infrared rays by the user's fingers the photodiodes send the videoscreen coordinates to the computer for processing.

A touch-sensitive terminal has four major components:

1. the color video monitor

2. an optical array of infrared light-emitting diodes and photodetectors

3. a protective plastic cover, and

4. a microprocessor touch controller linking the terminal to the computer.

In one of the available units, the matrix is composed of 64 infrared light emitting diodes positioned at the bottom and 48 along the left edge of the optical array. An equal number of photodetectors is located opposite the diodes along the top and right edges.

To accurately determine when and where the user is touching, the microprocessor in the touch controller monitors each horizontal and vertical beam individually in sequence, with all beam positions scanned about every 1/15 of a second.

The processor senses the amount of light at the photodetector just before the opposite light emitting diode is turned on and stores this information. Then it tunes the infrared diode opposite the detector, senses any changes in light level, and finally compares the two light levels.

The photodiode touch detectors have a resolution of about one-eighth of an inch. If this procedure detects a significant increase in light, the controller has the evidence that a finger has not blocked the light at that position. No significant increase in light indicates that the light beam has been interrupted.

The controller microprocessor computes the exact coordinates at which the beam has been interrupted, and sends this reference to the computer for processing. This coordinate information is used to determine the programmed sequence to be displayed on the color video monitor.

The application of this technology can range from banking systems to any type of information provision, including tourist information. In one implementation, the video information provided to queries comes from optical video disc players. Each video disc contains 54,000 separate video color frames. In addition to a minicomputer, the terminal controller contains special graphics display and musical sound synthesizers.

A block diagram of this computers and communications facility is given in Figure 13.2. The synthesizer provides audio responses to touch requests, and also background music. If the user asks to speak to an attendant, the terminal controller requests the central control function to perform the video, voice, and data switching thus completing the two-way video connection.

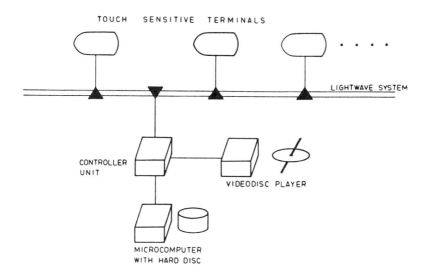

Figure 13.2 A PC based Interactive Videodisc system.

The application in reference has been designed by the Bell Telephone Laboratories and is supported by a specially developed high level computer language for programming the videodiscs. The programs are referred to as *scripts*. A video dictionary relates alphanumeric names to frame numbers on the video disc, while a flowchart debugging capability enables the script writer to correlate the video dictionary with the video disc.

Bell Labs has also been engineering a language which permits programming directly on the television screen, thus making it possible for the script writer to *finger paint* sections of the screen that are to be sensitive to the user's touch. An interactive software allows the script writer to create the whole set of alphanumeric characters and symbols used in the graphic overlays. The operating system is Unix.

The WorldKey Information Service at Epcot Center, Florida, consists of 29 guest touch-sensitive terminals with up to eight terminals connected to a guest terminal controller. The latter supplies the audio and visual information about the center.

Special requests are answered by attendants at seven touch-sensitive terminals. The network controller switches video and audio communications between a guest and an attendant and also updates information stored in the guest and attendant terminal controller. All video, audio, and data information is transmitted over a lightwave local area network.

13.5   OPTIMIZING DISPLAY CHARACTERISTICS

We have spoken of the wisdom of optimizing the man-machine interaction in a way of making it simple, friendly, and effective for the end user. Using closely related technologies has great potential—but it should be approached within the forementioned perspective.

One of the important display characteristics is the ability to dissociate the scratch pad function from menu selection; and menu selection from reporting. Another interesting possibility is the simultaneous presentation of tabular data and of the corresponding graphs.

Such requirements suggest the ability of using multiple videos—or of high resolution, large row x column dimension, flat panels with windowing capability. Classical low pixel video screens should not be effectively used for multiple windowing, because the resulting spaces are too restricted to support a valuable presentation of text, data, and graphics.

An area of interest lies in the development of flat panel displays. But this display would have to offer price/performance in the same range as future CRT oriented products with similar capabilities.

These capabilities would have to include the display of at least one standard page in less than one second at a resolution of about 100 lines per inch. The primary advantages expected from such a device would be portability and reduced desk space (smaller footprint) requirements. Windowing, smooth scrolling, and lack of jittering are musts.

A non-portable office version of a flat panel display would be expected to offer good capability. One of the primary requirements is the need to display at least two full pages of text simultaneously. Color is expected to be a requirement on many of the devices sold. Plasma and thin-film electroluminescent (TFEL) offer the greatest potential in terms of associated technologies.

While color would be a must, experience with PRESTEL and other systems indicates that in business applications no more than four colors (high contrast) should be displayed at the same time. These should be chosen from one of the evolving industry standard sets.

The type of technology is evidently important. There is general consensus that LCDs are too slow (20 to 30 ms) for most user interests. However, they offer good contrast, are visible in daylight (depend on reflected light), require low driving voltages and power, and may offer static or permanent memory (power off state) in the future.

Plasma (gas-discharge) displays appear to be increasingly common. However, the large number of connections and associated driving logic is keeping the price high for displays of significant size and resolution (in excess of $1,000). Smaller displays have been available at reasonable cost for some time.

Plasma displays allow resolution in the order of 120 lines/inch, a 60 Hz update rate, and reasonable size (Sony has a 1024 x 512 array). Disadvantages are:

- the relatively high cost
- high driving voltages (180 to 230 V)
- limited size/resolution
- relatively low brightness levels (10 to 25 fL)
- the monochrome limitation
- lack of brightness control

Thin-film electroluminescent (TFEL) displays appear to offer the most in terms of cost/effectiveness. The major disadvantages are that driving frequencies in the order of 125 to 400 Hz and voltages of 30 to 150 V are required.

Advantages include the potential for color, brightness levels in the order of 100 to 1000 fL (sunlight viewability), and ease of fabrication. This technology is largely within the lab but seems to be developing quickly with some electroluminescent displays already produced.

Vacuum-fluorescent displays (VFD) with a phosphor on anode/grid wire approach also offer some interesting possibilities. Amongst these are acceptable resolution (0.1 mm spacing); high brightness levels (80 x 150 fL); and reasonable power requirements (5.7 W for a 256 x 512 pixel array). Color is also possible with the use of multiple phosphors.

Chapter 14

VOICE INPUT/OUTPUT

14.1 INTRODUCTION

Integrated circuits have helped in voice handling from voice recognition and answerback to the improvement of the quality of sound, and also the development of intelligent telephone lines. It is, therefore, only reasonable that interest in business implementation is growing.

Speech technology is looked upon as the next major breakthrough in computers. While this approach is still in its infancy, its implications in banking, manufacturing, retailing, and transportation are widely recognized. Within the next 10 years, knowledgeable observers agree, verbal interaction with computers will be the rule rather than the exception.

There is a variety of opinions regarding various voice related technologies and their potential value. Some are in direct contradiction with each other, while the following points appear to have reasonable support:

1. Voice recognition would be *required* for certain situations. Both speaker dependent and independent discrete speech support would be needed. In the beginning, these solutions would be important in hands and/or eyes busy environments, but later they will be generalized. Vocabularies of at least a hundred words and speaker independence would be necessary for data capture. Voice transcription would be very important and require connected speech and vocabularies in excess of 3,000 words.

2. Voice prints could be used for systems access security. Speaker verification positively identifies a particular speaker by his unique voice print. Verification technology compares the spectral voice qualities of an utterance with a stored voice template to differentiate between people speaking the same words.

3. Voice store and forward capability will become more important and more common. Voice comments on text (memos) should be stored or associated with the text, thus creating an annotation facility. Compression is necessary to make effective use of digitized voice; and message broadcast capability is required. Voice quality should be natural but does not have be high fidelity.

4. The telephone need not be physically integrated into the workstation but should receive extensive support from it.

Dial-out and auto-answer capability should exist. The phone should be viewed as the I/O device for voice messages and even control. *Vocoding* is the realtime transmission of a verbal conversation from a speaker to a listener.

One way to cut the cost of local and long distance calls is to institute an Electronic Mail system. Another is to employ voice store and forward, which does not require executives to learn to use a keyboard. Store and forward voice message systems not only cut phone costs, but, more importantly, they provide for quick and efficient communications.

Voice store and forward involves the realtime encoding, compression, and storage of a speech message for later retrieval. While voice response primarily involves static messages, voice store and forward deals with continually changing messages.

In the general case, voice input by means of automatic speech recognition equipment, offers major potential for improving user/operator productivity. It is the only input technique which does not require the direct use of hands and eyes.

Voice input can replace or complement keyboards, function keys, tablets and other types of input devices typically employed for entering commands, alpha or numeric data. In the past, data gathering choices have been dominated by keyboards. Complicated keystroke sequences have often been required to evoke appropriate computer controlled functionality.

Today, a number of advanced human interface techniques can be used to improve both source data capture and the selection of appropriate computer controlled operations. These techniques, including voice input/output, are becoming an integral part of many applications.

14.2 THE VOICE HANDLING EFFORT

Automatic recognition of voice has been a major research goal since the early 1960s. At the time, the Advanced Research Projects Agency spent $20 million on the SUR (Speech Understanding Research) program for continuous speech recognition.

More important than a standalone voice facility is the integration of voice with the digital signal handling environment. This may take place in different levels of sophistication:

1. All digital transmission. Already many voice calls over the telephone wires get digitized and are transmitted over digital telephone trunks. The next step is to have the analog/digital conversion at the handset level so that it is digital signal handling end-to-end.

2. Voicedatabase search and retrieval. This extends communications into the databasing level. Present-day applications examples center around store and

forward. The next step is to have the digital voice messages manipulated through user-actuated digital deletion, editing, and annotation. (Wang has equipment of this type.)

3. Voice recognition and voice synthesis. Though these are the two ends of a man-machine communication they stand at different levels of research, development, and implementation. Speech recognition is not as advanced as speech synthesis. Also, within this class exist specific goals: discrete word recognizers, connected word recognizers, speaker-dependent, and speaker-independent systems.

4. Integration of voice and digital services. This goes well beyond speech synthesis and voice recognition into the area of text, data, image, and voice integration—not only for storage and retrieval but also to produce a meaningful end product. Some videotex systems work along this line.

Let's now look into what is available in speech recognition.

Commercial speech-recognition systems first appeared in industry in the early 1970s, used in demanding situations by people who have to tackle several tasks at once: aircraft pilots, quality inspectors, factory supervisors, warehouse employees, and the disabled are a few examples.

One of the more gratifying applications has been with voice controlled wheelchairs and prosthetic devices such as robotic arms. Voice recognition systems are used in

- warehouses and production floors to enter inventory and production data

- shops to instruct numerically controlled machines

- by quality control inspectors in some factories to verbally report their results to a speech recognizing computer as they carry out their work

Yet, while these references are real, even successful commercial systems are limited in terms of vocabulary. For over a decade, the problem of making a machine recognize normal, conversational speech by any speaker is not yet at the level of a generalized solution.

Currently available voice recognition systems work best with the speakers who trained them. They recognize small vocabularies of isolated words, typically from 40 to 200.

Difficult for computers to follow is conversational speech because it is fluid, fast, and irregular. The exact sounds a speaker produces depend upon the size of the vocal tract, how it changes as the tongue and lips move, and whether the vocal cords vibrate.

When we talk, we tend to produce many meaningless sounds, such as breathing noises, sighs, and coughs—while background noise tends to mix with ordinary speech.

Training computers to understand human speech is a challenging task. Instead of assuming at the design level which aspects of the sound signal are most important, it lets the machine determine this from speech that was read into it.

• When spoken to, the computer calculates the probability that the current signal would represent various possible utterances.

• It then picks the most probable choice, according to the frequency of occurrence of word sequences in its memory and the probability that a particular sequence of words would generate the observed sound signal.

In one test, after the system was trained, it correctly identified 93% of the words in new sentences. Its errors were often plausible substitutions.

At Texas Instruments and the Bell Laboratories, among other centers, researchers test machines with different approaches. As semiconductor prices drop, such devices may turn up in games, appliances, automobiles, and personal computers. Since voice-synthesis chips were developed a few years ago, computers have been talking to us. Talking back to some of them is still a more demanding, but not an elusive undertaking.

The rate of progress can be gauged by the promise of a voice-input typewriter by 1984/85 by Hitachi. Although this is only a promise, it *does* indicate the state of the art.

A listening typewriter is a potentially valuable aid in composing letters, memos, and documents. It might become a revolutionary office tool, just as the typewriter, telephone, and computer have been. With a listening typewriter, an author could dictate a letter, memo, or report.

What he says would be automatically recognized and displayed in front of him. The listening typewriter would combine the best features of dictating— such as rapid human output—and the best features of writing, such as visual

No human typist would be required, and no delay would occur between the time an author creates a letter and when he gets back the draft or final copy. But several technical problems must still be solved.

One of the most difficult technical problems for automatic speech recognition today is word segmentation. The boundary cues marking the end of one word and the beginning of the next, although good enough for human perception, are generally not clear enough for automatic speech recognition. Hence, most speech recognition devices commercially available today require the user to speak in isolated words.

A number of technical developments will be necessary until products like the listening typewriter reach marketability. Among them are the possibility to operate in the presence of background conversation or machine noise and the need for fewer training trials for a given level of accuracy.

Other technical references are the ability to use low-cost microphones, which do not need be precisely positioned, rather than expensive, noise-canceling microphones; and the achievement of a 99% plus level of accuracy. Still another is the design of low cost but polyvalent terminals. They have to be simple, intelligent, and mobile.

Also, they must be able to frontend simple or complex applications, to effectively handle the serial nature of voice messages, respond to the necessary input/output volume, and since the drafting board reflect the fact that human voice adds an extra dimension in communications.

This field of voice handling is broader than the casual observer would think. Among the applications areas we distinguish: voice coding for the modem market, static voice messaging, interactive retrieval using voice, factory floor data collection, other office functions, voice mail with dictation, order processing of all types, transaction processing initiated with voice, and a variety of voice response systems.

14.3 PROGRESS IN VOICE RECOGNITION

Voice recognition systems can be divided into two major categories: speaker dependent and speaker independent.

1. *Speaker dependent voice recognition* is the most common methodology. It operates by comparing the spoken word to a vocabulary entered by the user. The operator first trains the system by speaking each desired vocabulary word. A digital copy, or template, of each word is stored in the vocabulary, and when a word is spoken during operation, a digital copy of the trial word is compared with each stored template. The system selects the best fit and makes the selection available for use.

2. *Speaker independent voice recognition* is a more desirable approach, feasible only for limited vocabularies and requires a large statistical database for each word. This differs from speaker dependent recognition in that it replaces training by the user with a set of statistically sampled pronouncements of a particular word by a large and varied population base.

Users may operate the system by speaking the designated word commands without prior, individual training. Further advances in computer and voice technology are needed before speaker independent voice recognition becomes practical.

Voice recognition systems can also be categorized by the ability to recognize single words or complete phrases: discrete or continuous speech.

Complete phrases or continuous speech implies processing at the rate of about 145 words per minute. Single words or discrete speech assumes a distinguishable pause between words.

Furthermore, voice recognition equipment may be trained or untrained. Trained means the ability to teach the device to recognize a particular word, pronunciation, voice, or some combination of the three.

Current products mostly fall into the speaker dependent discrete speech category. These offer only limited vocabularies and have a high rate of error.

Progress has been slow for a variety of reasons. Some of the problems that have to be contended with are:

1. Large vocabularies. A literate person may have an understanding of up to 100,000 words and have a speaking vocabulary of 3,000 to 8,000 words.

2. The length and type of words or phrases in the vocabulary.

3. The presence of homonyms and other syntax issues.

4. The relatively high rate of information that must be processed (145 words/ minute).

5. The slurring or smearing of words in connected speech.

6. The alteration of phoneme values (sounds) by those preceding and following.

7. Variation in voice resulting from age, sex, geographic background, current emotional state, and so on.

8. The relatively high storage cost associated with a single word.

The response to these problems has been to limit and structure the vocabulary, allow only discrete speech, use multiple repetitions of a word to establish variance on templates, and provide feedback mechanisms during actual usage to verify input. Voice recognition technology generally assumes three stages of functionality.

• The first is *endpoint detection*. This is the ability to locate the end of a word and distinguish a word from background noises.

• The second is *spectral analysis*. This involves a determination of the frequencies present and the relative energy distribution across those frequencies. Part and parcel are digitizing and the application of signal analysis techniques.

• The third is *pattern matching*. This is a process of normalizing along the time scale and comparing with stored templates. The templates themselves will have been prepared in a somewhat similar manner in the case where the device can be trained and may include statistically developed variances in energy and frequency distributions and also in time.

Applications have been evolving over a timespan. Functions range from receiving touch tone input from a telephone (or attached keyboard) to automatically answering a telephone call by selecting and sending a spoken response over the telephone line.

We will return to these issues when we talk of voice response. At the voice input end, the application of this technology also spans many areas. At the current state of the art it is particularly of value in four cases:

1. When both eyes and hands must be used for other purposes than text or data entry.

2. When a higher rate of input is required than can be achieved with the hands alone.

3. Where the user is unskilled or unable to use other types of input devices.

4. When training costs are a critical factor.

Examples of potential use are in automatic language translators, direct data entry, and (as stated) voice actuated typewriters. Such typewriters enter into the broader perspective of the office area where both the original dictation and verbal editing and routing of voice mail will become available.

Among the areas where this technology is already being used are:

• Banking applications such as credit card number validation and banking transactions via the home phone.

• Voice programming of numerical control devices.

• Instruction support in automated flight training systems.

• The control of weapon and other on board systems in military aircraft.

• Devices for the handicapped.

Some PBX manufacturers market units which allow voice input of numbers and the words "yes" and "no." This is intended to provide easy access to wide private telephone networks without operator intervention.

The customer inputs an identification number for billing purposes plus the telephone number he wishes to reach. The system repeats the numbers back, receives vocal verification from him, and then makes the connection.

One of the recent implementations worth mentioning, voice communication on the Renault 11, exhibits a 128 K RAM, rate of 1 kBit per second (KBPS), and speaking time of about 2 minutes. Research aims to bring the rate to 0.8 KBPS. Another project, by connecting 16 devices together, reaches a vocabulary of 1,600 words—more than the average citizen is using.

These are only two of the examples which help document that speech recognition is emerging as an important interface technology. Speech input can reduce the amount of attention the user has to spend on the mechanics of recording information and allows users to concentrate on their primary task. Benefits include:

• reduced user training time

• increased worker productivity

• reduced secondary key input

• improved timeliness and accuracy of information made available through voice

Voice recognition devices provide added security. Texas Instruments (TI) and AT&T plan to use them to address the phone set for automatic call-up. TI and Renault project a "start motor" voice recognition. If anyone but the owner pronounces the words, the "his master's voice" auto triggers an alarm.

Portable analysis and synthesis speech (PASS) can also lead to a near revolution—as learning machines get into the act. It will also lead to greater productivity and/or safety.

- A visual message is interpreted in 6 min. by 75% of people.

- An auditory message, in 2 min. by 99% of people.

The examples which we have just considered enter both under the heading of voice input and that of voice output. We will be considering both of them.

14.4   FOCUSING ON VOICE INPUT

As stated, connected speech recognition technology is several years away. Hence, any meaningful discussion on voice input is largely confined to isolated word, or short phrase (utterance) recognition.

Terminals used by the general public or casual untrained operators, require speaker independent voice input. However, speaker independent hardware available today operates with only a very limited vocabulary.

Using menus and prompts, a vocabulary of the ten digits, yes/no, backup, and help would be sufficient. This type of voice input could be completely under control and resident in the terminal with the vocabulary stored in ROM. Host support is not required since the host need not know if the input came from a keyboard or voice input.

Some of the more sophisticated speaker independent voice input devices are adaptive. A standard set of templates is stored in ROM. As the speech session progresses, the terminal can be trained to the user's voice, and it becomes speaker dependent with a higher acceptance rate.

With terminals where the user is trained, speaker dependent voice input permits larger, more flexible vocabularies. This environment is less hostile since the employee wants to make the terminal work and cares about its performance.

In this case, the terminal firmware will control the voice input hardware, but the recognition vocabulary (digital templates) would be stored in RAM. This may call for downline loading when the operator signs on, and an upline dumping of the vocabulary after a training session (done offline), and at sign off time.

Vocabularies must then be stored on disc or some other media for recall at sign on time. An update capability is required.

If during operation, one or two words are consistently not recognized, the operator should be able to update that word or words without going to a training session of the whole vocabulary. This necessitates the upline store at sign off time.

Most voice input systems have a program controlled reject level or template match coefficient. If there is a long range change in background noise between the time of training and the time of usage, a supervisor may need access to adjust this level or coefficient at the terminal.

Voice input recognition accuracy can be greatly improved through:

- structured vocabulary

- the ability to dynamically change recognition ranges in a vocabulary, link a common vocabulary, and change vocabularies

Most systems available today have this capability and would need to operate interactively with the host application software.

Due to the severity of the consequences of a substitution error where the result could cause damage or injury, many systems will echo back the recognized input and not take action on that input until the operator says *yes*. If the operators says anything but yes, the system loops back to wait for input.

A reasonable and necessary extension to a voice input system is Speaker Verification and Identification. This refers to the ability of a system to determine that a person is who he says he is, or which of a set of possible speakers is actually speaking. This is accomplished by matching his voice patterns against a stored version of voice patterns for a person or group.

One example of a speaker verification system is at a TI computer center. It is used to control access to the main computer room. Through speech synthesis a person wishing to gain access to the computer room is requested to say various phrases which are in turn compared against stored voice patterns for that person.

Another critical component of a voice handling system is *Store and Forward*. This is a method to allow multiple persons to operate together: Information is stored by one person and forwarded when the recipient can receive it, or when the network can afford to send it.

A voice store and forward system normally consists of multiple voice I/O terminals, data compression hardware, and mass storage. The voice I/O terminal converts a speech waveform into digital data.

To accurately represent the speech waveform at least eight thousand 8-bit samples must be taken each second. With this data rate 1/2 megabyte of storage is required for each minute of speech.

There is a large amount of redundancy built into human speech. To reduce data storage requirements for a voice store and forward system, data compressors are used to remove as much of this redundant information as possible. Commercially available units combine speaker dependent recognition, and voice response and voice store and forward in a common base. For either voice response or store and forward, a speaker says a word or message. For voice response the local message is compressed in realtime and stored in the unit for future recall.

For store and forward, the message is compressed in realtime and transmitted via an RS232 port to another system for either immediate reconversion to voice or storage until the second person or system is ready to receive the message. The compressed bit rate is selectable allowing flexibility in choosing speech quality, memory requirements, and channel bandwidth.

A popular application for speech recognition is in Computer Aided Design (CAD). Calma offers a speech input option for their CAD systems, claiming time savings from 10% to 25%. This is a speaker dependent system with a vocabulary of 100 utterances.

CAD systems fit nicely with present speech recognition technology. CAD input stations represent a hands/eye busy environment where trained operators are involved exclusively. Also, a vocabulary of 100 to 200 words is sufficient for handling system commands.

Another application with similar technology requirements relates to assembly line/quality assurance workers. Again, trained employees are involved, and a limited vocabulary can be tolerated.

One of the largest users of speech recognition in the manufacturing process is Lockheed. (They have been using voice recognition in manufacturing since 1977.) With a total investment of $400,000 for 36 voice input terminals and minicomputers to run them, they report an annual savings of $650,000.

These savings come primarily from reduced cost of data collection. A different type of application for voice recognition is services where we call a database and input account number, or password, verbally—acquiring account balances, and so on.

A basic requirement for such a system is that it be speaker independent. It must also have a high recognition accuracy in order not to lose the confidence of the public.

The system should allow a potential network user to enter his identification number and the phone number he would like to reach. Speech synthesis should repeat this input to verify that he has been properly understood.

Another speech recognition application area is in consumer entertainment electronics. Available products include voice operated televisions from Toshiba and Sanyo, voice controlled toy cars, and so on. Toshiba and Sanyo each use speaker dependent recognition systems in their televisions.

A major problem with the use of voice recognition equipment is the degree of difficulty that will be experienced in conditioning or training the equipment and user. Most applications involve fairly sophisticated ease-of-use logic to deal with this issue.

Price and performance of the actual equipment do not appear to be worries. The current state-of-the-art allows for the recognition of one or two hundred discrete words as fast as they can be spoken. This is oriented to a specific operator. However, other user profiles can be downline loaded into board level products allowing for multiple users of the same equipment.

The next level of development would support connected speech and a vocabulary in excess of a thousand words. NEC claims to be able to support a couple hundred in connected mode with a relatively expensive device. In short, this level is not expected to be achieved in any cost-effective manner within the next few years.

Let's recapitulate the references which have been made in this and in the preceding sections. *Speech Recognition*, we said, refers to the act of having a computer determine what is being said by a human.

There are three categories in which speech recognition technologies can be organized. Included in all classes are both speaker independent and speaker dependent systems.

The difference between the two is solely whether or not prior training is necessary for the system to recognize a specific voice. Speaker dependent systems require training, speaker independent systems do not.

1. The simplest recognition solutions are in the class of *Utterance Recognition Systems*. These are capable of recognizing a limited set of words, phrases or numbers spoken discreetly, but with a pause before and after. Such systems are widely used where a training speaker is involved and a limited vocabulary (approximately 100 words) is acceptable. They are also used when speaker independent recognition is required.

2. *Connected Speech Recognition* is an extension of utterance recognition. Combinations of a set of utterances can be spoken naturally without imbedded pauses, separated by the recognizer. Pauses are still required between phrases because this separation cannot be done in realtime. The ability to recognize continuous speech eliminates all restrictions on the way information is voiced. This requires that the separation into words and recognition of those words be done in realtime. The constraint of such a system is that template storage is required for every word to be recognized. This limits the size of the vocabulary to a small subset of any language.

3. The most sophisticated category is *Phonemic Recognition*. This is a technique for eliminating the vocabulary limitations of the two previous classes, but the technical difficulties encountered in its implementation warrant its inclusion as a separate one.

14.5 VOICE RESPONSE

Voice response involves the creation of a template of spoken word or phrase of any length, the storage of this template in the system, and the reproduction of the voice message upon digital command. In earlier voice-response systems, recorded voices were used. Synthetic, digital voices have, however, become common.

Speech output is required in an eye busy environment, such as some assembly line applications. A supplement to or replacement of visual prompts, is another implementation area. Speech synthesis refers to the artificial generation of sounds (by computer) capable of being interpreted as speech by a human listener. There are primarily three categories of speech synthesis, each covering different types of applications:

1. The most basic form of speech synthesis is *digital recording* of a human speaker. While providing good fidelity (the speaker can be identified) it requires very high data rates (60 to 100 KBPS).

2. The second category of speech synthesis is known as *utterance encoding*.

With this method, words and phrases are digitized and then encoded using any of several data compression techniques to reduce the data rates required. Depending on the compression technique data rates from under 1,000 bits/second to a few thousand bits/second can be achieved. Fidelity can be good, but not as good as direct digitizing.

3. The third technique is *phonemic synthesis*, or synthesis by rule. This allows a virtually unlimited vocabulary since only phonemes are stored. Phonemes are the basic sound segments of speech. Approximately 60 are required to construct all the words of the English language. A set of rules is used to concatenate the phonemes into the utterances to be spoken. With the present technology this approach has the lowest fidelity of the three discussed.

Speech quality is affected by many factors, most significantly by the baffle and speaker selected to output the final sound. In terminals, this can be more of a problem than where to put the few chips required.

It makes more sense to provide an audio output plug on the terminal and offer, at extra cost, a choice of speakers. An earphone is useful when there are several terminals in one area or for privacy. The earphone would be very low cost, like small transistor radios use, and would be the personal property of each user.

Voice response systems typically require retrieval and assembly, a voice database, and voice communication. Voice output techniques are two main types:

1. *Waveform digitized speech*, also called freeze-dried. This is known for its high quality sound (the person's voice can be recognized), but the vocabulary is limited by memory size.

2. *Synthesized speech*, or phonemic generation. This is known for its flat mechanical sound, but the vocabulary is unlimited.

Waveform digitized speech is applicable with terminals where the user is untrained; an infrequent user, or the general public, requires high quality sounding speech. A limited vocabulary is probably acceptable since the terminal does a specific function and the speech is used to prompt and guide the operator through a specific task that seldom varies.

Synthesized phonemic speech is best for terminals where the user is trained. An employee who is paid to run the terminal can use synthesized phonemic speech since the operator can learn to understand the speech as quickly as he learns to run the terminal. The unlimited vocabulary allows longer instructions and prompts and permits more complex and varied tasks to be performed by the user.

Presently available are voice output boxes, board level products with RS232 interface, and chip sets. The easiest addition of voice output would be the RS232 box, or board, but it is the most costly since it ties up an I/O channel.

Most of the chip sets operate on an address bus, ready/busy interface. With waveform digitized speech the address is of a word or complete phrase, but in synthesized phonemic speech the address is of a phoneme.

There are also available speech synthesizers that accept an ASCII text string and use a speech-by-rule algorithm to generate a phoneme string. For words that cannot be handled by misspelling, the algorithm can be bypassed by an escape code, and the programmer has access to the phoneme addresses.

Waveform digitized speech in the terminal would be under complete control of the terminal firmware because of the limited vocabulary and specific functionality. Synthesized phonemic speech would be controlled by terminal firmware, but could also be accessed by the host program.

A speech-by-rule algorithm would be in the speech module so that the majority of messages would be ASCII text strings. The source of strings could be terminal memory or the host program.

The most visible application of speech synthesis is in toys and games. These devices take advantage of any of a number of speech synthesis chips available. Most use utterance encoding and therefore must rely on limited vocabularies.

A fast growing area for speech synthesis is in various warning systems. There is a broad range of warning systems using speech synthesis to convey the warning message. Present applications include warnings to elevator passengers, auto operators, aircraft pilots, and warnings from appliances: "too hot," "fire," etc . . . . For such applications the vocabulary could be kept quite small.

The choice of the synthesis system could depend largely on the severity of the warning, and therefore the intelligibility required. For example, if instructions are given in the case of a fire, there should be no ambiguity in what is being said. This could call for a digitized recording approach.

An application requiring a very large vocabulary is text-to-speech translation. Included in this area is a number of devices designed for the visually handicapped.

The coming years will see an increased use of voice response systems in the financial sector:

- balance inquiry
- payment systems
- interaccount transfers
- ordering cheque books
- credit authorization

At the same time, the voice mail and voice recognition markets are developing. Their usage in the financial industry will be oriented toward meeting the financial users' voice requirements, entering into home banking with a low cost terminal, establishing a base product easily extendable to other voice applications, and providing a leading edge customer involvement with this line of development.

The trend in voices related technology primarily consists of VLSI improvements which ease the implementation of designs developed in the last decade. Most notable are the numerous digital signal processing chips and chip sets, reducing to a chip the hardware necessary to implement transforms in realtime, while also reducing the cost of such hardware to under $100.

Chapter 15

DISC STORAGE

15.1 INTRODUCTION

One of the most significant effects of technological development has been the sharp cost reduction per bit of information on hard disc (HD) storage. This, in turn, has opened new perspectives on the utilization of hard disc for personal computing purposes.

In a broad, all inclusive reference, auxiliary memory storage devices for all types of computer gear can be classified in the following five categories:

1. Hard Discs

2. Add-on Memories

3. Flexible Discs

4. Magnetic Tapes

5. Optical Disc

The HD category can be further subdivided into classes of rigid devices:

- Fixed vs. Removable

- Large vs. Smaller Capacity

- Classical vs. Supercompact

The most advanced in terms of low cost and low error rate tend to be classical, very large capacity, Winchester type. Here the competition is very tough between American and Japanese companies. NEC claims to have achieved a bit error rate (BER) of 10^{-6} for the 3.5 GBytes—and projects the introduction of the 10^{-7} BER.

Kodak recently announced Isomax cobalt doping process for floppy discs. Though still in prototype stages, the new process could have implications for personal computer hardware architecture.

Isomax permits much greater recording density on storage media. Especially portable computers, could be built with one disc drive and even more storage capacity than is available now with two drives. Sharp drop in the cost per bit of information is one of the features—and the same is true of smaller dimensions.

The microdisc drives are coming. Sony and some American disc and disc drive makers agreed to a common 3½-inch disc standard, but other microdisc standard still remain, the 3¼-inch disc being the contender. Still, the standards battle underlines that the auxiliary storage units are being miniaturized.

Companies are also working on smaller size floppy discs: less than 4 inches in diameter which are expected eventually to hold the same amount of information as the now-standard, 5¼-inch diskettes. This dimension sees to it that the disc drive which operates the discs can be dramatically reduced. Together with compactness come lower power needs.

The drive to miniaturize the external memory started 10 years ago with the introduction of the sealed head/disc assembly (HDA, or Winchester). They represented a major switch in design philosophy.

At that time, most drives stored data on a removable discpack, with retractable read/write heads. When a pack was installed, the drive brought the discs up to speed and positioned the heads over the disc's storage area.

The removable discpack was itself a development over the fixed disc. It offered the advantage of interchangeability, eliminated file backup, and by being stored away it increased the storage capacity available to the computer system.

15.2   WINCHESTER TECHNOLOGY

The search for more disc capacity has been constant since disc storage was introduced to the market in the late 1950s. Two factors determine a disc's storage potential:

1. The number of tracks per surface or track density.

2. The number of flux reversals per inch, linear density.

This is the most critical along the innermost track; as tracks decrease in length toward the center, each should store the same amount of data. Therefore, linear density will be greatest along the innermost track. Track density is affected by the precision of the mechanism used to position the read/write head over the desired track, and the width of the gap in the read/write head.

Linear density is determined by the flying height of the head over the disc surface, the properties of the disc's magnetic coating, and the read/write head's gap length. A basic reason behind the importance of Winchester technology is that it has resulted in a steady increase in linear density from 5.6 kilobits per inch (KBPI) in the early 1970s to about 12 KBPI today.

Several factors make the Winchester more attractive:

1. They are more reliable. The mean time between failures (MTBF) for Winchesters is 8,000 hours, contrasted with 2,000 to 3,000 hours for removable discs.

2. They cost less. A 160 MBy device is typically going for $5,500 (quantity 100). The same capacity removable pack drives average $10,000. Greater bit density on the former is one reason for this difference.

3. Less space is required for the Winchester. They can be rack mounted rather than using the standard cabinets that house the pack drives.

4. Conversion to Winchesters is relatively simple. Even a gradual conversion with a mix of machines on one system is possible; and the software is available. Table 15.1 summarizes the improvements obtained with Winchester discs as contrasted to the removable disc technology.

TABLE 15.1 Improvements with Winchester discs

- 5.6 kbpi in early 1970s
- 12 kbpi today

Winchester vs. removable discs:

	Fixed	*Removable*
mtbf	8,000 hours	2,000 to 3,000 hours
Cost of 160 mby	$4,500	$9,000
Required space	Rack mounted	Separate cabinets

The introduction of the 8-inch drives, started a trend toward smaller, cheaper Winchester solutions. It also led to diversity in disc dimensions, and therefore to a renewed interest of standardization.

The market now encompasses different sizes of drives. The 14-inch discs store 160 MBy to 1 GBy. They are aimed at users of mainframes and minicomputers. The next class (in terms of dimension) is the 8-inch Winchester able to store 5 to 70 MBy. Some of these hard discs fit into the same space as the 8″ floppy disc drives.

The third class is the 5¼-inch Winchester storing 6 to 20 MBy. In this, as in all other cases, the density is steadily increasing. Some 5¼″ prototypes reach up to 100 MBy. Table 15.2 contrasts the 1981 sales statistics to the estimated 1986 market share of the devices under consideration.

TABLE 15.2 The hard disc market share

	1981	EST. 1986
Fixed disc 5¼″	16.4%	43%
Fixed disc 8″	18.6%	21%
Fixed disc 14″	22.5%	4%
Cartridge 5¼″	—	7%
Cartridge 8″	—	11%
Cartridge 14″	11.7%	8%
Discpack 8″	—	3%
Discpack 14″	30.8%	3%
	100%	100%

The fourth class encompasses hard disc drives less than four inches in diameter. They are aimed at intelligent WP and portable computers. It is this class which, as stated in the beginning, has not yet been standardized.

Diameter is a basic reference when we talk of "classical" HD. The dimension varies with time as new, smaller diameters characterize the supercompact. In this manner:

- *1982* has been the year of 5¼"—which is now subjected to fierce pricing.

- *1984* has been the year of the under 4".

While the dimension of the compact disc is increasingly smaller, elder dimensions characterize new families of storage devices. In this manner:

- The 14" moves toward higher capacity.

- The 8" takes the 14" place in the "established" capacity market for mini and midi.

- The 5¼" substitutes for the 8" on Desktop machines.

- The 3¼" orients itself towards the portable and the word processing (WP) market.

At the same time, the elder, and therefore, larger dimensions increasingly feature removable media. Removable discs for 3¼", for instance, are not yet ready.

- The 14" drives are increasingly used for mass storage. NEC, for instance, announces a 680 MBy by 8 aggregate at more than 5 GBy per Drive Box.

- In the 8"–10.5" domain, the trend is to implement densities of more than 200 MBy—up to 500 MBy for current technology.

- The 5¼" HD presents two varieties in function of height.

- The 3½" high drive features about 36 manufacturers and an estimated 142,000 North American shipments. The current range in density is 30 to 140 MBy.

- The 5¼" also comes in ½" high packages. These hold 6 to 12 MBy—but solutions involving 30 to 60 MBy are now in the laboratory.

- Finally, work on the less than 4" HD drives goes on both on removable and fixed media. The exact dimension is not standardized.

Sony has introduced a 3¼" drive. CDC features "Cricket," a 90 mm or 3½" drive which in 1983 will support 6 MBy; then 12 MBy and by 1984 25 MBy. For the first two densities, the Access Time is 85 ms. For the 25 MBy, it is 50 ms.

The *Transfer Rate* stands at 5 MHz 5 for all versions, or about 600 KBy/sec. This means a high speed cable. The same transfer rate is valid for the 5¼" drives, but for 8″ it is 15 MHz with access time now at 30 ms and going to 20 ms.

Other technological advances as well characterize developments in discs. The year 1982 saw the evolution of thin metallic media as contrasted to the current oxide coating. With thin metallic, the object is higher density. As an indicative reference oxide permits 20 MBit/sq.inch and thin metallic, 50 MBit/sq.inch.

Still another reference is the type of recording. Vertical recording (originally to be applied on floppy rather than hard discs) may bring the density at the level of 5 MBy per floppy:

- with horizontal recording we can obtain less than 1 MBit/sq.inch
- with vertical, about 3 MBit/sq.inch

There are today nine different solutions, of which two follow ANSI and one presents an approach with local intelligence. Among the reasons for differences, and the associated trends, we distinguish:

1. The drive toward higher density
2. The need for error correction
3. More DBMS integration at the subsystem level
4. Faster transfer rates
5. The use of Parallel Channels
6. Lower Line Count (e.g. parallelism at a 2-byte level)
7. A greater visibility into the subsystem

In spite of reasons for divergence, the modular design of drive interfaces can bridge from past to future architectures.

It is, as well, correct to take notice that different HD markets exist for different computers. For instance, current projections indicate that for CAD/CAM work-stations, favored dimensions for removable discs will be the less than 4″; 5¼″ and 8″. For fixed discs, the 8″ and 14″. Correspondingly, for:

- PC, the fixed discs will come at the less than 4″ and 5¼″ dimensions
- Portable, ≤ 4, 5¼″—and the same for terminals
- Small business system, ≤ 4, 5¼″
- WS standalone, ≤ 4, 5¼″

A major problem with all types of hard discs is the introduction of intelligence at drive level. This involves:

- Software routines (drivers)
- Logic over data (LOD) and
- Error handling

The introduction of intelligence can impact quite significantly on design characteristics, particularly at the lower range of computers. The change of an HD in a personal computer might impact its design very much in terms of functioning—and this is not admissible as it affects a whole range of tightly coupled devices.

15.3   REMOVABLE DISCS

While removable discs have been around for more than 30 years, a persistent problem is that no two manufacturers make drives that can read the disc of the other. This is so even if both use the ANSI physical standard.

Stated in a different sense, the ANSI standard gives mechanical compatibility, but from the reading head, to the sector format, and logical recording the solutions are different and hence, incompatible.

A similar statement can be made of interface trends.

- While for telecommunications and voice/image handling may be used all diameters.

The explosion in the usage of hard discs should not be interpreted only through the statistics on a head-count but also in terms of storage capacity. In this manner, while the North-American utilization statistics for the 14″ discs indicate a growth in number of units from 360,000 to 400,000 which is rather stable, the installed capacity increases significantly. In 1981 there were on 14″ HD 4.4 Billion Bytes. For 1986 the estimate stands at 8.4 Billion Bytes.

Furthermore, by 1986 the 500 MBy HB will represent the 45% of installed total vs. 7% in 1981. Another major trend for the 14″ is toward fixed discs. From 35% in 1981, to 80% in 1986.

The following statistics reflect on the other dimensions:

- For the 8″ disc, from 150 K Units in 1981 to 750 K Units in 1986

This, too, is valid for the North-American market. Revenues will develop from less than $1 Billion in 1981 to $6.8 Billion in 1986. Let's also notice that in 1981 some 60% of sales was at the 5 to 30 MBy level, while in 1986 over 70% will probably be at 30 to 100 MBy.

More significantly, for the 8″ discs, in 1981 nearly the 100% was fixed. This is expected to drop to 70% in 1986 with balance divided between removable and dual HD.

• In the market for 5¼″ the development looks like from practically zero in 1981 to 2 million units in 1986—at about $7 billion.

Table 15.3 presents the estimated market for the less than 4″ discs in North-America and Worldwide. Other technical trends regarding the fixed disc are shown in Table 15.4.

TABLE 15.3 Estimated market for the less than 4″ HD (in thousands of units)

	1984	1985	1986
North America	320	800	1,360
Worldwide	484	1,280	2,430

TABLE 15.4 Trends relative to the fixed disc (circa 1984)

Access Time	16–20 ms
Transfer Rate	15–24 MHz
Area Density	24 MBits/39 in.

Intelligent Interfaces with about 16 KBytes

Removable Level	160 MBy
Dual Level	33 + 66 MBy

A very significant tendency is cost reduction. For fixed disc technology, cost by 1985 will probably stand at the $1,000 level for 5¼″, 36 MBy units; and $650 for 3¼″, 30 MBy (transfer price).

Streamer tapes for disc backup are expected to remain relatively expensive. The Streamer in 1985 will probably sell at $1,000—which suggests that the current cost difference between the 14″ HD and the Magnetic Tape units tends to remain the same.

Should such pricing trends prevail with a nearly equal cost for disc and tape, the pull toward disc usage will continue. In terms of access time and other technical characteristics, the user is much better off with discs. Furthermore, by the end of 1985 at about equal cost the user will obtain 140 MBy on disc vs. only 60 MBy on tape. What is further needed to help materialize a full disc-oriented orientation is the development of the appropriate software for the simultaneous update of concurrent, synchronized database images.

15.4 FLEXIBLE DISCS

A different species altogether are the flexible discs (floppies, diskettes). Originally developed by IBM in the early 1970s as a program loading device for big computers, they have since become a popular storage medium for small engines because of their low cost.

The first floppy solutions stored data on discs eight inches in diameter. In 1976, Shugart introduced a smaller and cheaper drive that used a 5¼-inch disc. This type found wide use in desktop systems.

Now, a new generation of smaller drives using 3½-inch discs begins to appear. Though their target is portable computers, they will find a significantly larger area of application.

As it has happened with the hard discs, the diskettes are progressing in terms of storage capacity. Early solutions provided relatively modest storage support. They used frequency modulation (FM) to encode data on one side of a disc at a track density of 48 tracks per inch and a linear density of about 2.8 KBPI.

In this manner, the 8-inch drives stored 77 tracks per surface. The 5¼-inch drive used about half that number. Storage capacities were roughly:

- 400 KBy on an 8-inch disc
- 125 KBy on a 5¼-inch disc

To meet the goal of an increased storage capacity, a modified FM (MFM) recording scheme was employed. It made it feasible to double the bit density (storage capacity) of both the 8″ and 5¼″ discs.

Then closed-loop servo systems have been employed that sense track position and thus enable the head actuator to follow changes caused by media distortion. This permitted to further increase the density of recording. (The information required to make such adjustments is stored as data on the track itself in the production process.)

But increased density brings along with it other technical problems. It requires increasing the data transfer rate of the drive, and this affects the complexity and cost of the drive controller.

A higher track density calls for a reduced track width, making increasingly difficult for the head positioning mechanism to stay on track. This is particularly challenging as the disc's mylar substrate tends to expand and contract with heat or humidity changes.

Yet, in spite of the greater potential for off-track errors, manufacturers have succeeded in doubling track density on both 8″ and 5¼″ disc drives. Shugart and Tandon market 5¼-inch drives that store a megabyte of data at a 96-track-per-inch density.

Though home computers, particularly those for games, can use the floppy disc as a storage medium for programs and data, this practice should definitely be discouraged with the professional PC. At the workstation level, the valid solutions are microfiles on hard disc, and LAN for communication among WS and with the local database.

Yet, floppy discs will be around for a period of time, until semiconductor prices break even with diskette storage—and the latter are replaced through cassettes with solid state software.

The two major functions to be done through floppies today, particularly for standalone PC, are program loading and bootstrapping. A *bootstrap* is a small program that gets the computer up and running every time we turn it on.

Advice on how to handle the floppy, as long as it is still around, starts with a DON'T: Never insert any disc before the computer is powered on. Never attempt to power down the machine until all discs are removed.

In fact, the first action necessary is to *Configure* or *Customize* the program on the floppy to our equipment. (This is normally done by the dealer who sells the the program but if it is not done, the Configurator Notes available with the program should be of help.

After the disc has been configured for our equipment and we have made a backup copy of the disc in case of failure, we should insert the configured disc into the primary drive and boot or reset our system. Well-written programs, when loaded on the PC system, will display the general functions or dispatcher menu.

The floppy disc, or diskette, is a thin, round circle of magnetic recording material within a lightweight square of plastic support. The disc is made of the same material as recording tape and it stores information in much the same way. The logical part is more challenging.

Very important, though not appreciated as they should be, are the disc save and restore procedures. Every magnetic medium will someday fail. The diskette surface will eventually wear out or be damaged (scratched). So we should be very careful to backup our important documents. For instance: keep an extra copy on a separate disc.

Magnetic discs require special care in order to keep them in good shape. The user should never touch the discs recording surface. He should handle the disc by the plastic covering, and be careful not to touch the exposed parts of the disc surface.

It is good practice to always return the disc to its protective jacket when it is not in use. And since most disc drives read and write on the back side of the disc, we should *not* lay them down face up without the jacket on them.

Environmental conditions are also important. We should not expose the disc to extreme heat, direct sunlight, or magnetic fields. Scissors, paper clips, and other metal items may be magnetic and they will erase or change information on the disc.

At the disc drive, always insert the disc so that the label faces the door of the disc drive. Furthermore, follow the instructions of the vendor. Many vendors give written advice—which should be carefully considered.

15.5 VERTICAL RECORDING

A new means of recording has therefore been sought, able to serve both hard and flexible discs. Developed in Japan, *vertical recording* is still in its beginnings. Discs with vertical recording could be introduced in 1983, though volume production is projected for the mid-1980s.

Vertical recording proponents argue that it will pack more bits on a surface than optical recording: Vertical recording can lay down 100 KBPI as contrasted with the 25 KBPI of optical recording.

In its fundamentals, the vertical recording technology works as follows: As horizontal, or longitudinal, magnetic recording attempts to squeeze magnets lengthwise, vertical recording squeezes the magnets' thickness. Both flux strength and flux rate of change are affected.

The squeeze of the magnets' thickness improves storage capacity by increasing the magnetic domain densities. The same process makes the data bits easier to detect. In the electrical signal generated in the magnetic domain, 0, 1 storage is represented by polarity reversals in vertical recording; by peaks in horizontal recording. (Figure 15.1).

Vertical recording is promising an increased storage capacity as it circumvents a density limit imposed by the horizontal recording. Technology can make able use of the electrical signal waveform in the read/write head.

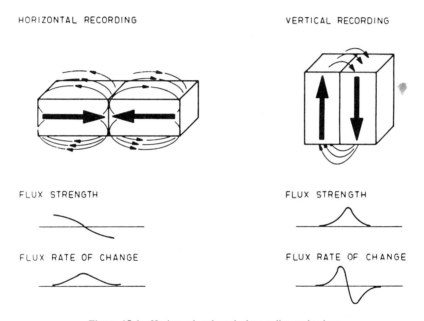

Figure 15.1 Horizontal and vertical recording technology.

Each peak represents a data bit. As bit density increases, the domains decrease in size until the signal waveform becomes too small to be detected. With current heads, this point occurs around 20,000 to 25,000 flux changes per inch (KFCPI). Domain size is independent of bit density, flux transitions are sharper than in longitudinal recording, and error rates are much better controlled in high recording densities.

15.6 STREAMING TAPE FOR DISC BACKUP

Disc drives with removable tape cartridges provide automatic backup for fixed disc media in PC and LAN. That saves users from buying separate streamers or regular magnetic-tape units.

The *streaming tapes* are fast moving, readily accessible tape units with enough varieties to keep users confused. Most of these tape cartridges are designed to fit a special niche of tape storage.

- One is the 14-inch streaming tape drive for disc backup.
- Another is the 12-inch streaming cartridge tape drive.

Streaming tapes are beginning to be used with Personal Computers, some mini, CAD/CAM and Local Area Networks as a backup facility to hard disc storage. Six tendencies will characterize development and subsequent employment during the next couple of years:

1. Reduction in mechanical complexity
2. Data density at 20 KBPI (vs. the current 8 to 10 KBPI)
3. Cache buffering
4. Advances in Coding
5. Chromium dioxide media
6. Lower cost

A reduction in mechanical complexity is necessary not only for the present mainly journaling options but also for the introduction of a start/stop capability. The principal object of Cache (at the 4 KBy level) is for error correcting. Another objective is higher performance.

Streamer solutions will impact on the use of disc storage. With some offerings, a disc and a tape can be run on the same controller unit and share a common buffer (for instance, in multiples of 4 KBy shared as a flip-flop by discs and tape. In turn, this will influence file design, presenting an interest in keeping the files at the 4 KBy level.)

Advances in coding will increase the streamer's storage capacity—a necessary feature as the discs the tape intends to backup become increasingly potent. Media such as chromium dioxide are projected both to improve dependability and to cut down on cost.

Furthermore, streamer solutions introduce delays particularly associated with dumping operations: The following are the key activities:

1. Insert Streamer
2. Copy on disc
3. Rewind Tape

As fixed disc backup devices, streaming tapes come in two dimensions which exhibit strong growth characteristics: ½" and ¼".

Both are in cartridge. One of the major problems among available technical solutions is data interchange. At the present time there are no standards and, therefore, no interchangeability.

It is possible to program the machine to dump the transaction file on a pre-established basis—for instance, every hour. This calls for the writing of an automatic scheduling program.

Special software can also see to it that a shadow image is recorded on a second disc. As an alternative to the streamer tape this calls for a program to do the dual recording and another to control the synchronization.

Table 15.5 presents statistics in terms of capacity, recording algorithm, data transfer rate, access time, and positioning time. These statistics are relative to a new ¼″ announcement "Sentinel" by CPI/CDC. The streamer features 11 tracks, 450 feet in length, read as a serpentine with 0.7″ gaps between blocks.

TABLE 15.5 Statistics on streaming tape SENTINEL

1. Data Capacity:
    42 MB with 2 KByte blocks
    48 MB with 4 KByte blocks

2. Recording Mode:
    Interface   NRZ
    Tape   NRZ I

3. Data Transfer Rate, average over block:
    440 KBPS or
    55 KByPS

4. Tape Velocity: 55 inch/sec

5. Access Time: 61 ms

6. Positioning Time: 290 ms
    Hence, repositioning   351 ms

The repositioning reference (351 ms) is valid in case the tape is used in a start/stop mode. If the blocks along any one of the tracks are read in a streaming fashion, then the access time is 5.4 ms with an interblock gap of 0.6″. Change of tracks (along the 11 tracks of the ¼″ tape) does, however, involve a delay of 740 ms.

The following statistics reflect the Mean Time Between Failures (MTBF):

- For first six months of production: 6,000 hours.
- Between the six months and 18 months: 7,500 hours.
- After the 18 months period: 10,000 hours.

Most manufacturers foresee a market explosion for streaming tapes: from some 200 units sold in 1982 to about 32,000 units in 1986—these references concern the North American market.

Let's add that with the advent of the streaming tape the market now features five alternatives:

  1. Vacuum Column drives/vertical

  2. Vacuum column drives/horizontal

Both are developments of the classical tape drive, the latter being the newer and often preferred version. These tape drives are for professional data processing centers.

  3. Cassette drives.

Small music cassettes are widely used with PC for home and hobby applications. The drive can be a tape recorder.

  4. Video Cassette recorders.

Some PC and LAN use VCR backup, some operating in an analog fashion, others in digital.

  5. Streaming tape drives.

Oriented, as discussed, to the professional PC, CAD/CAM, LAN markets, but also to the home market.

On a cost basis, the 12-inch streaming tape provides 20 minutes of dump-restore function to back up Winchesters, and sells for $995 in large OEM quantities.

15.7   OPTICAL DISCS

Optical discs have a feature that even the best tapes cannot match: random access. Yet, few manufacturers have shown interest for data storage, as opposed to visual-image storage.

The reason is the error rate. Low cost entertainment optical discs have error rates on the order of one in 10^6 or 10^7 bits.

- With video devices, the eye is forgiving.

- But if we drop data, as optical discs do, we are in a lot of trouble.

Error-correction schemes are expensive, though resourceful designers will continue to improve the discs.

Professional optical discs achieve a bit error rate (BER) of one error in 10^{14}. This exceeds by over two orders of magnitude BER characterizing current magnetic BER rigid disc media.

For read only purposes, the optical disc advantage is its huge capacity potential for low cost.

Optical fiber technology is in its infancy, but can offer bandwidths of several GHz (1.0000 MBPS). This reference is particularly important to solutions regarding carrier technology, at a cost that is currently two to three times greater per meter than that of coaxial cables. And these costs will probably come down.

Since labor cabling costs are the same per meter, current cost differential between optical fiber and coaxial narrows. Connectors to these higher bandwidth cables, however, are much more complex and costly.

Coaxial cable connectors require a microprocessor-controlled buffer memory and repeater to eliminate data corruption that results from current surges on the connected terminal. Optical fiber cable connectors are even more complex. They need to include optical to electronic and electronic to optical signal convertors.

Signal conversion is one of the major problems with optical recording as well—and so is the fact of high bandwidth, which tends to justify the higher conversion costs. Because increased disc capacity by a hundredfold outpaces the cost increases; there is hope for many technical problems to be solved. One is that of not being able to erase what has been written, because in an optical system a laser beam burns tiny holes in the coating of a disc (up to a potential 10 billion holes, each one a thousandth of a millimeter in diameter, on a 14-inch disc).

It may be possible soon to overcome that difficulty, but until this takes place optical storage seems best suited for archival data, with magnetic discs complementing the optical recordings. Several companies are experimenting with laser optical technology: IBM, Storage Technologies, Magnavox, Xerox, RCA, Philips, Toshiba, Hitachi, Thomson CSF—and results should be forthcoming.

Research on tellurium as a recording medium is going on in several centers. The medium is critical, for the holes must open up uniformly on writing, so that when the disc is read, light passes easily through the holes and the transparent substrate.

Initially, optical discs would be useful for big databases that are permanent or semipermanent. In this frame of reference they will find competition from both magnetic discs and tapes. IBM is rumored to be working on an eighteen track tape drive with massive amounts of storage, probably in the gigabyte range, for disc backup.

Optical discs will also find usage in new areas such as computer aided instruction (CAI). At least three manufacturers, IBM, NCR and DEC, are announcing CAI systems with color display, the images being stored on optical disc and the variable text/data on magnetic media.

Cost-wise, it is projected that the first disc (master copy) will carry a high price tag (some $50 thousand) but subsequent copies will cost $20 or less. This makes the process ideal for mass markets, eventually integrating on the video disc digital audio (voice) capabilities.

For archiving purposes, juke-box type solutions are expected around 1987—along with a new range of applications. In terms of implementation, optical disc technology might prove to be quite similar to the experience we have had with bubble memories: It never got where the experts thought it was going to be and finally it dropped off the market.

Chapter 16

BENCHMARKING THE PC

16.1 INTRODUCTION

At the vendor's side, a polyvalent strategy is necessary in planning for new products. Such strategy typically involves design, marketing, and pricing characteristics. As a senior General Electric executive was to suggest: "The next generation of Apple products will make the difference whether the company prospers or disappears." And the same is true of all other PC firms.

As far as the PC manufacturer is concerned, three basic issues should characterize the new product line of companies working in the frontiers of knowledge:

1. The able use of high technology for improved performance keeping the options open for future development.

2. The sharp knife to cut costs and keep the product competitive.

3. The ability to invade the competitors' market and conquer their strongholds.

Issue No. 3 can be served through this advice: Be able to play a *first class game* within the business world, master a plan to attack the competitor's own market, but also assure compatibility with your own installations to avoid software upheavals. Hence, be able to play in two frames of references at the same time.

For the computer user, the goals are quite different than those characterizing a manufacturer's strategy. A computer-enriched workplace is an investment. Return on investment will come by way of:

• Improved productivity—both mental and clerical

• A better quality of work

• Return on investment by way of the productivity improvements

Since return of investment takes into account both ways and equipment costs, the lower the latter are the lesser the time to break even. It is useless to expand capacity at the PC level beyond a certain level answering the local processing, databasing, datacomm, and end user requirements at the workbench.

To apply this concept the right way, we should first evaluate what the needs are—then benchmark the equipment to assure that it responds in an able manner to these needs. This is the objective of the case study we will be following in this chapter.

16.2 THE RANGE OF MACHINES BEING CONSIDERED

The goal of working on the data supported by the Benchmarks is to extract information which permits one to evaluate:

1. Computer makes available in the market; for instance, the IBM PC, Olivetti M20, two Apples (II and III), the Hewlett-Packard 125, Atari 800, Corvus Concept, and Sinclair's ZX 81 (the latter for control purposes only). As we will see, with the exception of the Sinclair machine, the other equipment did not present an inordinate variability.

2. A methodology uniform enough to provide the basis for the coming evaluations. For any benchmark, the methodology to be followed should be most carefully chosen, but once this is done, the methodology being selected should be observed in the unavoidable additional tests to be made—for the results to be comparable. The same is valid of the types of tests which have been conducted.

3. The identification of significant differences which prevail among programming languages employed for precisely the same processing algorithms. As we will see, the language test further documents the wisdom of selecting Pascal, over Basic and Cobol. Let's now look at the type of tests and the obtained experimental results.

Eleven simple tests have been selected for the Benchmark. They are identified in Table 16.1, and represent operations which show up quite frequently in applications.

With the exception of operations No. 10 and 11, which are more of an engineering kind, the others constitute vital parts of a business type computer implementation. To these have been added file creation examples as we will see in subsequent tables.

The turnaround time of the seven PC in relation to these 11 simple tests is given in Table 16.1. Such tests range from loop to string concatenation and they have been executed a number of times—as properly identified in the table's footnote.

The programming language for each machine is BASIC. If, to the seven machines identified in the Table, we add Corvus Concept—with the reservation that it has been programmed in Pascal, and Pascal executes faster than Basic—we are led to the conclusion that six of the machines cluster rather closely

TABLE 16.1 Benchmark on seven PC
(with programs in BASIC)

	M20	IBM PC	Apple II	Apple III	HP 125	Atari 800	ZX81
* 1. Loop for next	8"2	8"9	10"2	13"3	13"8	21"	2'34"2
* 2. Add integer	26"	31"8	31"3	34"5	38"4	51"	5'20"5
* 3. Add fraction	23"2	32"1	32"3	35"4	39"5	51"	5'22"1
* 4. Multipl. integer	27"	33"8	40"5	40"5	40"2	1'11"	6'00"4
* 5. Multipl. fraction	27"1	35"2	50"3	44"9	46"2	1'28"	6'06"
* 6. Division integer		43"5	45"3	42"2		1'41"	
* 7. Division fraction	28"1	43"6	54"1	47"8	56"5	2'30"	6'25"6
**** 8. String concaten.	3"2	3"2	13"5	19"0	13"2	4"7	15"78
*** 9. Matrix manip.	4"2	4"2	55"0	46"0	13"2	6"	32"7
**10. Square root	7"5	9"4	48"5	34"0	28"8	2'19"	7'52"4
**11. Logarithm	9"0	9"7	22"2	16"5	15"5	2'09"	5'01"8

*10,000 times
**1,000 times
***20 × 12: repeated 3 times.
****255 characters, added one at a time, repeated twice.

together. The ratio from the best to the worst is, in many cases, 1:2. Two machines are outsiders:

- The Sinclair ZX 81 performs miserably—and constitutes an example of a home computer for the kids.

- The Corvus Concept gives the most outstanding results—an excellent case of a professional computer.

It will be quite interesting to establish, during a subsequent Benchmark, whether other engines based on Motorola's 68.000 give results like Corvus.

A similar statement can be made regarding the test results of the IBM PC and those to be obtained by other engines with Intel 8086 microprocessor. (Allowance should, however, be given for the fact that the current IBM PC uses an 8088 processor. Between 8088 and 8086 there is a 30% processing difference because of the 16-bit bus of the latter.)

Figure 16.1 brings in perspective the results obtained by the tests in reference from three different engines:

- Apple II
- Apple III
- Olivetti M20

With the exception of the file creation and engineering type operations, the difference among the three engines is not significant, though the M20 performs consistently better than the other two machines. Most important, the Apple II/ Apple III results definitely mediate against the latter. This is more expensive without a corresponding benefit.

Table 16.2 is interesting in the sense that the Atari 400 was a $70 engine. With the exception of square root and logarithm, at execution time the turnaround ratio to the much more expensive IBM PC varies between 1.43 (for matrix manipulation) and 3.44 (for fraction division).

For the same tests which have been outlined, Table 16.3 contrasts Apple II, Apple III, and Concept—programmed in Pascal. On two occasions (tests No. 1 and No. 7) in the Apple III column are given two turnaround times. The second number reflects the use of Apple III without video, which allows the machine to operate with a 2 MHz clock.

It will be appreciated that the difference is below 20% for Apple III, while in the overall Apple II and Apple III were not significantly different in a way to justify the higher costs of the latter.

In terms of obtained results, the difference is really impressive between the two Apple engines and Corvus Concept. As a general trend, there is an order of magnitude between the turnaround of the Apple engines and that of the Concept. This reference is valid for the 11 test programs and also for the Pascal compilation.

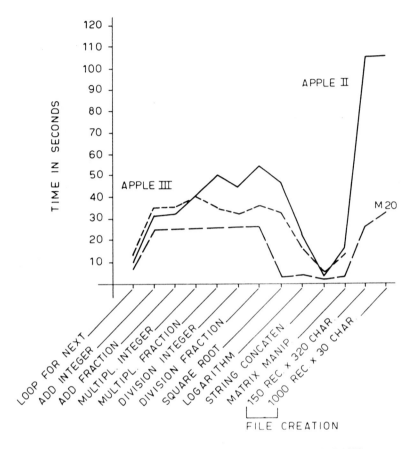

Figure 16.1   Benchmark Apple II—Apple III—M20 programs in BASIC

Figure 16.2 brings such results under perspective. Programmed in Pascal:

• Apple III presents a steady lead (albeit a minor one) over Apple II in the business operations—but fails in the mathematical, as it presented problems with these tests.

• Apple II demonstrates a dreadful performance with file creation—particularly in the case of the 1,000 records of 30 characters each.

• The Corvus Concept excels in all tests: from the simple business type, to the engineering and the file manipulation.

Significant differences in turnaround have also been found between the Corvus Concept and the IBM PC (respectively programmed in Pascal and Basic). Table 16.4 demonstrates that the IBM PC is slower by a factor which varies between 1.74 (division of fraction) and 11.05 (matrix manipulation).

TABLE 16.2 Benchmark on IBM and Atari
(with programs in BASIC)

Program	IBM PC	Atari 400	Ratio 400 PC
* 1. Loop for next	8″9	21″	2.36
* 2. Add integer	31″8	51″	1.60
* 3. Add fraction	32″1	51″	1.59
* 4. Multipl. integer	33″8	1′11″	2.10
* 5. Multipl. fraction	35″2	1′28″	2.50
* 6. Division integer	43″5	1′41″	2.32
* 7. Division fraction	43″6	2′30″	3.44
**** 8. String concaten.	3″2	4″7	1.47
*** 9. Matrix manip.	4″2	6″	1.43
**10. Square root	9″4	2′19″	14.78
**11. Logarithm	9″7	2′09″	13.30

*10,000 times
**1,000 times
***20 × 12; repeated 3 times
****255 characters, added 1 at a time, repeated twice.

TABLE 16.3 Benchmark on Apple and Concept
(with programs in *Pascal*)

Program	Apple II	Apple III	Concept
* 1. Loop for next	6″11	4″26 3″51	0″20
* 2. Add integer	9″20	6″48	0″34
* 3. Add fraction	19″40	14″75	1″96
* 4. Multipl. integer	15″50	10″79	0″50
* 5. Multipl. fraction	33″12	24″25	2″
* 6. Division integer	9″95	6″99	0″50
* 7. Division fraction	45″83	34″13 28″44	2″50
**** 8. String concaten.	36″97		3″07
*** 9. Matrix manip.	38″00		3″50
**10. Square root	—	—	—
**11. Logarithm	2″54		0″38

*10,000 times
**1,000 times
***20 × 12, repeated 3 times
****255 characters, added 1 at a time, repeated twice

Notes

(1) Tests on Concept were made only in Pascal; No BASIC available.

(2) Turnaround without Video, hence 2 MHz. With Video a III works at 1 MHz.

(3) A III presented problems with these operations.

Finally, the time of Pascal compilation was

11″ with Concept

2′00 with Apple III

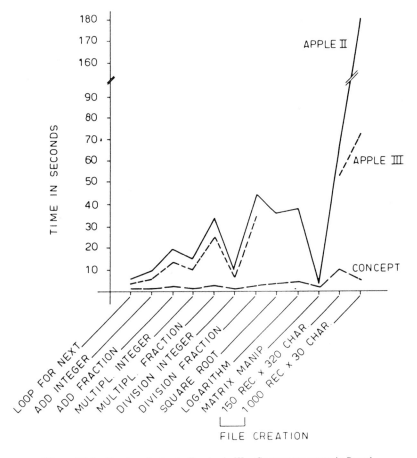

Figure 16.2   Benchmark: Apple II—Apple III—Concept programs in Pascal

Just as significant are the results which have been obtained in benchmarking the same machine: Apple II, with Pascal and Basic. The execution of precisely the same applications is slower with Basic by a factor which varies between 1.18 (for the division of fraction) and 6.45 (for matrix manipulation). Table 16.5, however, indicates a glaring exception: the computation of logarithm is faster with Basic than with Pascal.

In Table 16.6, similar results point to Pascal as the much more efficient language on Apple II. Figure 16.3 brings such differences under perspective.

Finally, it is worth taking notice that Apple II and Apple III are quite different in the operation of file creation (Table 16.7). Still, in this important issue the winning engine is Concept with differences ranging between 6 and 12 times when the Pascal language is being used.

The same file creation operations have been tested with the Basic language on Apple II and M20 (Table 16.8). The performance of M20 is more than three

TABLE 16.4 Benchmark on IBM and Concept

Programs	Programs in BASIC IBM PC	Programs in Pascal Concept	Ratio IBM/ Concept
* 1. Loop for next	8″9	0″20	4.45
* 2. Add integer	31″8	0″34	9.35
* 3. Add fraction	32″8	1″96	1.63
* 4. Multipl. integer	33″8	0″50	6.76
* 5. Multipl. fraction	35″2	2″	1.76
* 6. Division integer	43″5	0″50	8.72
* 7. Division fraction	43″6	2″50	1.74
**** 8. String concaten.	3″2	—	—
*** 9. Matrix manip.	4″2	0″38	11.05
**10. Square root	9″4	3″07	3.06
**11. Logarithm	9″7	3″50	2.77

*10,000 times
**1,000
***20 × 12; repeated 3 times
****255 characters, added 1 at a time, repeated twice.

times better than that of Apple II. But if the M20 is compared with the Corvus Concept, this ratio turns around against M20 with the M 68.000 based engine being the winner.

This can be stated in conclusion regarding the bench marking tests: When we talk of PC choice, turnaround time is one of the most fundamental characteristics of online systems in general and of interactive systems in particular. When the

TABLE 16.5 Benchmark BASIC-Pascal on Apple II

Program	BASIC	Pascal	Ratio BASIC/ Pascal
* 1. Loop for next	10″21	6″11	1.67
* 2. Add integer	31″33	9″20	3.40
* 3. Add fraction	32″28	19″40	1.66
* 4. Multipl. integer	40″50	15″50	2.61
* 5. Multipl. fraction	50″27	33″12	1.52
* 6. Division integer	45″30	9″95	4.55
* 7. Division fraction	54″12	45″83	1.18
8. String concaten.	—	—	—
*** 9. Matrix manip.	16″39	2″54	6.45
**10. Square root	48″49	36″97	1.31
**11. Logarithm	22″18	38″00	0.58

*10,000 times
**1,000 times
***20 × 12, repeated 3 times

**TABLE 16.6 Benchmark BASIC-Pascal
on Apple III**

Programs	BASIC	Pascal	Ratio BASIC/ Pascal
	13"36	4"26	3.14
1. Loop for next	10"51 (*)	3"51(*)	2.99
2. Add integer	34"50	6"48	5.32
3. Add fraction	35"44	14"75	2.40
4. Multipl. integer	40"50	10"79	3.75
5. Multipl. fraction	44"89	24"25	1.85
6. Division integer	42"20	6"99	6.04
	47"78	34"13	1.40
7. Division fraction	39"37 (*)	28.44 (*)	1.38

*Turnaround without Video, hence 2 MHz. With Video A III Works at 1 MHz.

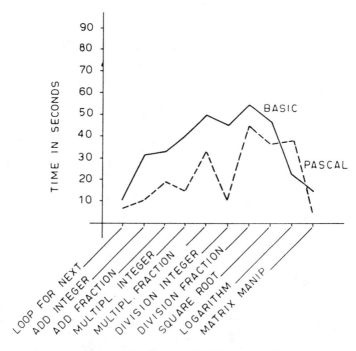

Figure 16.3 Pascal vs. BASIC on Apple II

TABLE 16.7 Benchmark on file creation
Apple II, Apple III, Concept with Pascal

	Apple II	Apple III	Concept
Files with	1'07"10 (*)	52"58 (*)	9"47 (***)
150 rec. × 320 char.	1'09"27 (**)		
File with	3'01"10 (*)	1'10"99 (*)	5"55 (***)
1,000 rec. × 30 char.	2'18"24 (**)	1'02"11 (****)	

*Registration done on a floppy of 5".
**Registration done on a Nestar LAN with only 1 WS active.
***Registration done on an Omninet LAN with only 1 WS active.
****In this registration (on floppy 5") the Video was not on.

response time is slow, the whole servicing operation is delayed with the result to increase the clerical costs as well as the unhappiness of the customers being served.

16.3   THE CHOICE OF EQUIPMENT

Benchmarks offer a factual basis for equipment selection, helping the prospective user to:

1. Focus on performance in an objective, documented manner, which is supported by test results.

In the last analysis, performance criteria will impact on cost/effectiveness and, therefore, constitute a critical basis for selection. But while performance is the first reference, it is not the only one. Without any doubt, the second criteria in the evaluation should be:

2. The ability of the PC under evaluation to support a standard OS. In order of preference:

TABLE 16.8 Benchmark on file creation
Apple II, M20 with BASIC

	Apple II	M 20
File with		
150 rec. × 320 char.	1'45"	27"
File with		
1,000 rec. × 30 char.	1'45"	32"

- MS/DOS
- Unix
- CPM

and with it a standard language.

The evident reference is the Pascal language. Esoteric OS and languages should definitely be excluded from consideration—and this is just as valid of the machines which have nothing but colloquial software support. Another basic criterion is:

3. The libraries of routines available with the PC under consideration.

A fourth fundamental criterion is:

4. The maintenance network available for the support of the PC to be chosen.

Users are being advised from an increasingly wide range of sources to orient themselves toward self-maintenance. Yet, the manufacturer's ability to support its own equipment, also providing users and third party maintenance firms with the needed spaces and knowhow, remains a critical factor. A similar reference should be made to:

5. The manufacturer's ability to demonstrate advanced knowhow able to guide the user's hand toward the benefits derived from the implementation of advanced technology.

Another fundamental feature whose existence should weight on the choice is:

6. The availability of a local area network architecture, and that of communications protocols.

Quite evidently:

7. Cost is a basic consideration in terms of choices and should, therefore, be brought under perspective in establishing the source of procurement and the configuration to be chosen.

The reference to cost is further underlined by the IBM PC XT Personal Computer with up to 20 MBy of disc memory.

Let's not lose from sight that at regular intervals IBM announced significant price cuts to the currently available PC and its attachments. The machine with 256 KBy, keyboard, floppy drives, color graphics monitor, and graphics printer now sells at a very competitive price.

Commenting on this significant price cut, mandated by technology, IBM's general manager for Operations was to remark: "It was necessary to assure that we maintain our position in the marketplace."

Price cuts are a good reference and should be observed by all PC manufacturers to be called in the competition for the commercial network. In fact, my proposal is to benchmark all computers in terms of cost/effectiveness against the first IBM PC for the DP chores, and the second IBM PC XT for the integrated hard disc capability.

In the last analysis it is irrelevant if we take IBM's computer or any other— provided that the forementioned seven conditions are met. What is relevant is that the cost-effectiveness of any other PC is significantly higher than that of IBM to make up for the relative risk taken in adopting the wares of any other manufacturer.

16.4 IMPACT OF THE OPERATING SYSTEM

Great care must be taken in planning the benchmark. Test results will only then be valid when the projected operating environment has been properly reflected in the tests. This has been the reason for including the File Tests.

Machine capacity is only one of the four key factors which impact on work-station performance. The other three are:

- The Operating System

- The language with which the applications are written

- The system design of the applications themselves

Table 16.9 compares Pascal vs. Cobol on a set of operations. These have been run on an esoteric OS (Mod 400) which is highly unadvisable for personal computers.

TABLE 16.9 Pascal vs. Cobol on 6/10 with LSI 6, Mod 400, 512 KBy, HD

	Cobol	Pascal	Ratio
1. Loop for next	39"10	<1"	39.1
2. Add floating point	1'32"70	<1"	3.7
3. Multiply floating point	3'32"70	28"	8.5
4. Divide floating point	3'54"00	41"	5.8
5. Matrix manipulation			
12 × 20	9"30	1"	9.3

It is easily observable that Pascal beats Cobol by a significant margin. Other tests confirmed this outcome. It is simply unwise to use an inefficient program-ming language at the WS level.

Table 16.10 compares the OS: MS DOS vs. Mod 400. Both tests have been programmed in Basic. The same equipment has been used. Here, performance under MS DOS exceeds that under Mod 400 by more than 1,000%.

TABLE 16.10 MS DOS vs. Mod 400 on 6/10, with LSI 6, 512 KBy, Basic Language

		Mod 400	MS DOS	Ratio
1. Loop for next	10,000 times	1'30"	6"30	14.0
2. Add floating point	10,000 times	4'21"	19"	13.3
3. MPY floating point	10,000 times	4'34"	20"	13.7
4. DIV floating point	10,000 times	4' 7"	20"	12.3
5. Square root	1,000 times	1'33"	2"	46.5
6. Logarithm	1,000 times	1'32"	4"	23.5
7. Concatenation 255 numbers	2 times	21"	1"	21.0
8. Matrix manipulation 12 × 20	3 times	25.2	1"	25.2

There is no way of beating an efficient, commercially available OS—such as MS DOS, CPM, and Unix—through the esoteric offerings by the microcomputer manufacturer.

• Not only the latter would tend to lock the user into the vendor's own box—which is deadly and should be definitely avoided.

• But also the resulting efficiency will be depressingly low.

In Table 16.9 (discarding the extreme value of loop for next) the average Cobol to Pascal ratio stands at about 7. This is conservative; other tests have given a Cobol to Pascal ratio of nearly 10. As far as the PC are concerned, Cobol is by nearly an order of magnitude less efficient than Pascal.

Table 16.10 suggests that the inefficiency of the esoteric Mod 400 (which, after all, was a minicomputer development) as compared to MS DOS standard at an average value of 2,100% (!). If we take this OS factor of 21 and multiply by the ratio of 7 for languages, we obtain 147.

In other terms, the results of Cobol on Mod 400 as contrasted to Pascal on MS DOS running on precisely the same type of equipment are separated, in the average, by more than two orders of magnitude. Cobol on Mod 400 is proven to be 1,470% *less* efficient than Pascal on MS DOS.

These are results no organization should discard for any reason. Though users will be welcome to develop their own tests, such tests must be properly structured and able to point toward a well-documented decision.

16.5 THE SYSTEM STUDY

The choice of equipment should be made with a specific system study in mind. This is necessary for dimensioning both the network and its moving gear in order to do the work in the most effective manner, and within the agreed upon time-tables.

The study of the workplace with the goal of a successful implementation will typically involve the following components:

1. *Fundamentals*

- functionality to be supported
- text and data rates
- text and data volumes
- interactive features
- response time
- reliability—uninterrupted service
- human engineering
- user friendly approaches

2. *Mechanics*

- programming language
- applications software
- visualization
- database structure
- database access (to be kept at minimum)
- authorization/authentication
- journaling
- recovery/restart
- possible add-on interfaces

3. *WS Support*

- growing range of OA routines
- DP/WP integration
- document handling
- transaction control
- form control
- electronic mail
- backup
- handholding

Most particularly, man-information interactivity calls for attention on:

- Response time (40% at 2 sec., 30% at 3 sec., 20% at 6 sec., 10% at 10 sec.).
- Uninterrupted service (99.99% uptime).
- Controlled access to microfiles.
- Softcopy presentation.

A systems view must be taken—and though its implementation may be timed, it must also be assured that subsequent development will fit nicely with what has been done.

Such capabilities can be implemented through a small, spirited team with a primary task modeling for an easy-to-use computer-supported management environment. This is a new philosophy in industrial environments—and it calls not only for advanced technology but also for a good deal of explaining and acquisition of knowhow.

By providing quicker, stronger, and more profitable relation to the clientele, online systems have become a major competitive edge. We cannot dismiss technological progress, but we can neither forget the drawbacks which will result if we don't design the system in the most careful manner.

A good example is the special attention to be given to response time. I take as an example an accounting application on a minicomputer with only six terminals where response time reached unacceptably long delay levels. The reasons thought to be in the background of this result can be classified as follows:

1. *Type of work (W)*. There exist eight alternatives:
W1: Voucher input—without tax update.
W2: Discounts, exceptions—without tax update.
W3: Accounts payable—with tax update.
W4: Discounts, exceptions on AP—with tax update.
W5: Accounts receivable—with tax update.
W6: Discounts, exceptions on AR—with tax update.
W7: Different accounting updates.
W8: General ledger updates.

2. *With and without overflow (A)*.
A1: With overflow (nonrestructured files).
A2: Without overflow.

3. *Online only vs. online and batch (B)*.
B1: Online only.
B2: Online and batch.

4. *Level of archive accumulation (C)*.
C1: Minimum level—early year.
C2: Maximum level—late year.

5. *Number of terminals (T)*.
T1: 12 terminals.
T2: 6 terminals.
T3: 3 terminals.

To test all conditions, a total number of $8 \times 2 \times 2 \times 2 \times 3 = 192$ cases would have been necessary. As a result, the experiment was divided in two parts. Conditions 1, 2, 3 were taken first.

Through the use of grecolatin squares, 16 rather than 32 experiments were done. Table 16.11 shows the average response time per experiment. Overflow and batch tend to significantly increase response time.

TABLE 16.11 Average response time per experiment

	A1,B1	A1,B2	A2,B2	A2,B1
W1	23″		2′00″	
W2		54″		19″
W3	27″		24″	
W4		1′18″		29″
W5	26″		24″	
W6		2′30″		28″
W7	3′58″		5′08″	
W8		2′00″		46″

Next the number of archive accumulation and number of terminals were tested:

• The increase from three to six terminals tended steadily to increase response time from 18″ to 30″—roughly by 66%.

• More impressive has been the inclusion of past files (late year emulation).

• For the same number of terminals, the average response time increased from 30″ to 2′34″—or by 513%.

The results of the experiment identified the system weaknesses which had to be corrected. As a result, a new system analysis was done which focused on these issues and reached quite satisfactory results.

1. The creation of a local area network with PC-based, intelligent workstations to be installed at each desk.

2. The reorganization of the database access mechanism, expected to improve significantly the overall performance, when historical data are needed.

The latter reference underlines the necessary attention to the availability of communications protocols.

Databases have locality—whether they are taken centralized or distributed. Even within the same program, the locality issue exists, and the problem is synchronization. This brings under perspective:

- *Access control* (which is relatively cheap)
- *Integrity control* (which can be expensive)
- *Design semantics:* that is, the ability to differentiate and to imply controls.

In a distributed environment there should be no accidental deletes and no propagation effects. Propagation effects confuse the design issue and impact on the integrity.

Among the critical parameters for file allocation algorithms in a distributed environment we distinguish:

1. *The level of data sharing.* It indicates whether the DBMS is partitioned or replicated, and to what extent.

2. *The nature of the access patterns,* which may range from inquiry only to total update, and can even vary significantly from time to time.

3. Whether the *information* on the nature of the access patterns is *deterministic or probabilistic.*

While for the starting conditions a valid estimate will be helpful, still more important is the monitoring of file utilization and the reallocation of files because of the dynamic nature of the interactive work.

Another problem to be carefully studied is workstation identification and replacement. When we replace something there should not be an update of, say, seven places—but only one affected. This, too, is part of the design semantics to be considered.

The distributed database, and the fact that it is widely accessible, poses prerequisites quite different than those of centralized processes. We must change our image of system design and adapt it to the new requirements.

We must also study the disc alternatives on which the database will be stored at each location. These will be selected in connection with the PC choices—yet each PC choice can offer one or more possibilities.

Table 16.12 identifies the alternatives currently available. While without doubt the preference is for hard disc of the smaller dimension, but higher capacity.

It is necessary to keep in mind the backup requirements. I expect the latter reference to be one of the most difficult issues to solve in connection with any system.

TABLE 16.12 Disc alternatives

HD	Floppy	Cartridge
14″	—	14″
8″	8″	8″
5¼″	5¼″	5¼″
*3½″	**3½″	***—

The cost ratio HD to cartridge is > 1:3, but might go down to 1:2.
Streaming tapes for backup are 14″, 12″ with a coming (CIIHB) 9½″. They run up to 48 MBy.

*The < 4″ HD is not standardized. Sony advances the 3½″; MPI/CDC the 3¼″ (Cricket). Starting at 5 MBy it will be 10 MBy in 1984, and 25 MBy in 1985.

**Both Sony and CDC advance this floppy.

***Irvin/Olivetti works on a 3½″ cartridge.

Part Four
THE COMMUNICATIONS WORLD

Chapter 17

EXPERTISE IN
TELECOMMUNICATIONS

17.1 INTRODUCTION

There are many professional fields where knowhow mortality is high, but none compares to the short lifecycles characterizing experience and specialization in telecommunications, in this decade.

This reference goes all the way from protocols to communicating databases, online workstations, and the use of wideband, and local area networks. Quite often, it is difficult for users to think in terms of new perspectives in communications because up to now most devices built for datacomm have been designed to operate voice grade telephone lines.

This frame of reference is changing rapidly. AT&T has made a statement that by 1990 the whole long lines (long haul) network in the United States will be digital or electronic analog. Today, this is true of about 60%. By the end of the century, the whole network would be digital in the sense of an end-to-end transmission. Long distance services will be characterized by three tendencies:

1. Single Side Band Radio

2. Fiber Optics

3. The New Generation of Satellites.

Fiber optics are considered to be already commercially viable—but the technology is not mature. In 1984, both AT&T and MCI work on fiber optics implementation in the Boston-New York-Washington path—and more developments are coming by way of fiber optics implementation. AT&T predicts that fiber optics will eventually become the medium for transmission, cabling, and networks.

The transatlantic cable will be fiber optics and supplement the satellites linkages. ''To be competitive, you need all three technologies,'' said a cognizant executive.

Video conferencing also has great potential. When full motion is not required, it can be replaced with freeze frame images which will gain in popularity.

Since managers spend so much of their time in meetings, AT&T anticipates a huge market for teleconferencing and sees this as its principal offering for the integrated office. The company expects to expand into networks and terminals and to have an integrated office product offering by 1985.

Just the same, significant focus will be placed on on-premises and corporate networks. AT&T's plan to move to a 64 kbps (kilobit per second) standard for transmission speed, and it comes as no surprise that it has developed a strategy for communicating databases and office automation.

At the same time, independent communications networks will move forward to establish a wide customer base. Families with simple PC as home computer terminals can hook into *The Source* and *CompuServe* through their telephone line, displaying information on their television screen or on a simple automatic printer.

The Source offers financial data, general news from United Press International, games, electronic mail, and even the right to "publish" articles electronically. CompuServe supports similar services, helping American newspapers test electronic home delivery in an experiment with the Associated Press.

This new trend to public communications attracts thousands of customers each month. At $5 to $25 per hour Americans are becoming skilled at the logical issues characterizing a modern communications utility.

17.2   TRANSMISSION SPEED AND TYPES OF TERMINALS

One of the important terminal characteristics of a communications system is the transmission speed which it supports. Traditionally, the most widespread standard speeds of the data transmission vary from 300 lps to 2.4 kbps (kilobit per second)*, with higher speeds up to 9.6 kbps and 64 kbps being also in demand. Between 300 bps and 64 kbps the ratio is 1:213.

Though, as we will see, the moving gear behind the transmission speed is the modem, the terminal too plays a key role in the issue of transmission speeds applied to a network. Given the capabilities (and limitations) of the human operator and of other factors, the speed of transmission is in direct relation to the possession of buffer memory. Such transit memory interfaces the high speed of the electronic equipment and transmission line from the far lower speeds of the operator and of the electromechanical devices of a terminal.

---

*Bits per second is a unit of transmission. It identifies how fast information is carried through the lines. One of the confusions often found in industry is that bps is used synonymously with "Baud."

Baud is a unit of modulation, a figure indicating the number of times per second that a signal changes (we will be talking about it in the proper chapter).

Although most of the terminals installed today operate at 1.2 kbps or less, the trend is definitely for terminals to work faster. This is particularly true in business, industrial, and financial applications. Higher transmission speeds than 1.2 kbps are advisable in case of heavy dataloads and for a better exploitation of the central computer resources and of the available line capacity.

A terminal is designed for a certain range of transmission, projecting the interfacing and datacomm work which it should be doing. To satisfy transmission requirements, not one but many component units must be properly designed. In particular, in the hardware side:

* the input/output devices
* the memory
* the physical interface(s)
* the modem

As will be underlined with reference to the modems, we ought to bear in mind that the physical interface of the terminal toward the modem must respect international standards. At the same time, in terms of software a terminal is characterized by its:

* protocols
* operating system (OS)
* text and data exchange packages
* local applications routines
* self-diagnostics

The logical interface is the linkage to the central system's software and to the other terminals or nodes in the transmission network. This gives rise to the question of compatible ways of connecting, an issue to which we will make reference when we talk of protocols. For the time being it is sufficient to remember that protocols and logical interfaces are a way of handshaking.

If all data processing equipment and transmission facilities were designed and built by one vendor, the relationships might have been by design, sufficiently interrelated to preclude a need for hardware/software interfaces. With communications, however, our data will necessarily utilize equipment from several vendors and operate over telephone lines that were never intended for data. Many of the problems which we are facing with datacomm arise from this simple fact.

Depending on the design characteristics which they project, terminals can be of varying types. Generally, they are grouped into five large categories, distinguishable by the presence of memory, calculating capability, and programmable features:

1. The lowest class includes simple question/answer, non-intelligent terminals, linked directly to a computer, whether mainframe or mini. Such terminals are teletype like and, with the appropriate design, can serve both telex/TWX lines and data communications, a main difference being the line speed. This class works start/stop (S/S, asynchronous) line discipline and typically has a one-character buffer able to hold, for example, the eight bits of a character.

2. Terminals endowed with a transit memory or buffer belong in the next class. Such a buffer will typically be able to hold several characters, either for parallel character transmission purposes, or to avail a more efficient interface. The buffer will allow the operator to work at his normal speed in introducing the message. It will also assure that the line is occupied for a very short time: the time needed for emptying and refilling the buffer—and not for transmitting character by character at the rate of message preparation, which is bound to the speed of the man.

3. The third class offers more facilities at the terminal point, through the dual approach of presenting a greater data storage capability *and* elementary programming features. Not only the printing or message preparation time can take place without occupying the line, but in addition, the terminal can be used for some local error correction, before transmitting the message to the computer. Or, it may automatically repeat the message in case of line errors, prompting the refusal of faulty messages, and the correct storage of those it has received.

4. The fourth category includes more complex data terminating equipment, with a programmable memory, able to perform complete operations in checking the accuracy of the message, hence reducing the number of duties to be carried out by the computer resource which acts as host. Data can be sent in a compact form from the computer to be treated by the terminal programs, prior to presentation to a human operator. Transmission time is thus reduced and so is the overall data traffic. Visual stimuli able to help the operator or end user, e.g. I/O check, ready, run, keyboard error, cancel, can assist a man-machine communication.

5. The top class consists of the most highly developed terminals, typically: microprocessor-based, endowed with memory for data and programs, and also storage devices: floppy discs, hard discs, and the like. They can operate both online and offline, in most cases being a personal computer.

The fifth class includes logical operations, the ability to augment the number of storage devices, and the use of micro files handled through local supports. Because of the highly reduced costs of the semiconductor components, the line between sophisticated terminals and microcomputers is hard to draw.

Intelligent terminals are the trend in peripherals technology. Design methodology, input/output processing techniques, programmable versus hardwired designs, special purpose applications, fault diagnostics and steady cost reduction are the key factors on which will be based the terminals design of the future.

But while the sophisticated terminals will increasingly enter the industrial and business scene, the simple, low cost devices will always have a place where rock-bottom economy rather than heavier duty is the main issue.

17.3 DATA CONTROLS AND SECURITY FACTORS

No matter which terminal a user chooses, he wants to be sure that it can support a data control faculty. Errors can come from a simple input (e.g. the operator working on the keyboard), from the equipment itself (machine errors), or, from the transmission line (usually, noise*), or further still, from another equipment communicating with *this* terminal.

The amount of loss of precision in a quantity; the difference between the original quantity and the transcribed or transmitted one, are examples of errors. Through hardware or software, a terminal must afford the needed controls to check and correct errors resulting from different sources. Let's examine the reference to those controls which specifically concern the input data by the operator.

The object of a *hash total*** is to check whether a list (block of data) has been correctly input by the operator, an error in keying-in is made, or a figure is lost in transmission. In the same class of controls falls the *check digit*, or parity. If to the number 12345 is added a check digit (say "odd/even" type: 1 + 3 + 5 = 9, minus 2 + 4 = 6, and 9 − 6 = 3) then the number will be written: 123453. The last digit "3" is for control purposes.

If during transmission the number is altered or the operator inputs by mistake: 132452, the machine will do the foregoing calculation, establish that 13245 does not correspond to the check digit " " and signal: ERROR. The ability of a terminal to respond to such occurrences is generally known as: visual input check.

The possibility for checking the input data, as it is keyed-in, must be complemented by that of a feedback to the operator. This is usually carried out through the printing of the text itself or through its visualization. This faculty is so important that, apart from very few exceptions, it is possessed by most terminals. More sophisticated error control may involve:

- the control of field* length
- the description of the omission of fields
- the assurance of the total introduction of a field

The latter examples relate to *Format*—that is, the way the data is organized and presented on hardcopy or through visualization. Such checks are applicable

---

*"Noise" is any unwanted input.

**A summation of field used for checking purposes, which has no other useful meaning.

*A set of one or more characters, not necessarily lying in the same word, which is treated as a whole; a unit of information.

both in regard to data input and the transmission/reception of messages. The utilization of terminals, which allows the control of data in transmission, is of great importance in a data communications environment. Format controls can be obtained by means of software. Other cases are:

• preventing an alphabetical field from invading a numerical zone (the assurance that in a numerical there are no alphabetic characters)

• determining whether or not a numerical zone be subject to parity checks or hash totals

• verifying a functional sequence

• normalizing a zone (the word "normalizing" should be interpreted with reference to the operations which follow and their requirements: for instance, data transmission, handling by another computer)

• carrying out functions such as printing, read/write on magnetic support, and the like

• performing insert/delete of specific fields of data into the records stored in memory

Control possibilities may exist at different levels of complexity. Their presence makes the terminals very different from one another. Therefore, it constitutes a parameter which must be considered most carefully in a comparative analysis.

We must also consider the management of the output data, generally known as Editing. It means to arrange information, the deletion of unwanted data, the selection of pertinent data, the insertion of information prior to printing, and zero suppression, etc.

Editing is mostly done through software: for intelligent terminals by their own program; for the others through commands transmitted online from a computer, minicomputer, or microcomputer. In either case, the aim is formalizing and directing the handling of data:

• editing the fields

• injecting (or, alternatively, suppressing) blancs, zeros, punctuation signs

• using service characters for the structuring of the printed (or visualized) output

• in case of printing, returning the carriage to the origin

• printing on different modules depending on the nature of the data

• varying the position of the printer on the printing lines, and

• controlling the paper feed (for instance, skipping/jumping lines)

A terminal which will allow the forementioned kinds of data management will be that much more useful to the man, or the organization, faced with specific editing problems. On the other hand, such terminals are more expensive.

Another important parameter to consider is the use of terminals accessible by authorized persons only. This facility can be effected through software or hardware, the former allowing a greater flexibility and sophistication than the latter.

Like a safety box, a banking terminal may be locked by two keys: the one is kept by its operator, the other by the manager of the branch office; both keys being necessary to put it into operation. Another terminal design—for instance: a cash dispenser—may accept a magnetic stripe identification*, but also for security purposes, prior to giving out money, it requires that the user keys-in his personal identification number (which will be different than the one imbedded on the magnetic card).

To compare the two numbers and reach an accept/reject decision, the terminal must either have its own storage device, or reach online into the memory of a computer. In both cases, this service is provided through software.

An extension of the authentication capability is the use of levels of priority. If many users have access to a computer network, management may choose that certain qualified persons can have different levels of priority than others. The same is true about machines and programs. Priorities make a system better able to respond to service requests when and where it is needed.

It is as much for safety (authentication) as for priority reasons that a terminal must, at every moment of operation, function well leaving no doubt to the user that the data transmitted and/or received is correct. The examples we have given, and many others concerning the functional characteristics of terminals, have the same reference point: the logical, programming capability of the information system interfaces which we call terminals.

Software, i.e., the programs written for a given terminal, can help provide a whole range of services: security controls, error checks, event sequences, fulfilling of tasks, format controls, temporary memorizing data for input/output purposes, forwarding data as instructed, correcting data before being sent online improving line usage, repeating automatically a faulty message, and so on. Many of these possibilities can have, directly or indirectly, to do with security factors. In other cases, the key issue is ease of service. Both safety and service are more vital to an industrial, financial, or business environment than to the hobby computer user.

17.4   THE USE OF TERMINALS

A terminal is designed to serve within an applications environment. The demands posed by this application identify the technical and operational requirements relative to its usage. The use of terminals brings in evidence the need for a classification in terms of applications. We can distinguish the following basic types:

---

*Typically, the usual credit card with an imbedded magnetic stripe.

1. Terminals for single transactions and inquiry. Typical applications along this frame of reference are Realtime and Timesharing. The user communicates with the machine with brief messages. He wants immediate access and needs a small amount of information in response.

The end user or specialized operator dials a message with the aim of updating files stored on a mainframe, minicomputer, or microcomputer. He may or may not wish to receive a reply. If yes, the reply can be presented on video or printed.

For this type of application the terminal can be extremely simple, like a teleprinter, or the class "1" identified in a preceding section. To keep costs low, both in absolute and per transaction, the printer's speed can also be low, say at about 30 or 40 characters per second (cps). But terminals used for this job can also attain a certain level of sophistication, such as a minimum of controls, a variety of conversational capabilities, a local treatment of data, and the like.

2. Terminals for data communications. This is an extension of Type 1, specifically designed for interactive dialogue with the computer to which it is connected or addresses through a line discipline. The word "extension" should be interpreted with specific orientation to more advanced data communication capabilities, for instance, packet switching. (Packet switching is a communications discipline which employs big blocks of data, for instance, 8,000 bit per block, as the basic unit of transmission.)

Such terminals must be able to transmit blocks of data from a remote periphery to some other point, on transmission lines. They must also be able to answer the physical constraints of man-machine communication, and provide not only a speedy transfer of data collected on the periphery to the center but also visualize the data received from the center for the periphery. Typically such terminals will be endowed with storage media to extend their central memory capabilities.

3. Terminals for multiple (Remote Batch) transactions. These are generally keyboard terminals with input/output oriented components. They are employed to collect a large quantity of information, using control procedures. The data is input onto the keyboard, stored in memory and converted to supports: floppy disc, streaming tape, etc. But it may also be entered through a floppy or other support.

Such terminals operate on-off. After the needed controls are made, the data is transferred in block to a remote computer (Remote Batch Processing). To handle this, calls of problems, they are in a position to receive blocks of information, transferred physically or online from the computer; as well to expedite blocks of data (file transfers) as the need arises.

4. Terminals with peripheral inputs.

Depending on the type of application and method of operation, a terminal may have access to a range of peripheral input devices, acting as concentrator besides its own work as an input unit. Similarly, a terminal can manage several output devices—for example, video units.

The important issue in "type 4" is the existence of terminals able to run other terminals, directing their operations and treating them like subsystems. This gives a different dimension to the work which can be done in the periphery. It also increases the terminal's cost, and further underlines the need for choosing the right terminal.

Since we said that a personal computer is the most versatile terminal available today, it is important to underline that the foregoing paragraphs in no way imply the use of multiple keyboards on a PC. Quite to the contrary, the personal computer should be a dedicated engine able to bring the power of the microprocessor to the end user's desk.

As the last reference brings under perspective, with applications techniques steadily developing, with the advance of technology, and with distributed systems and online solutions, the computer gets ever closer to the man. It is important that man-machine communications take place in simple and easy terms. While simplicity in usage must be increased, delays must be reduced. Waiting time has to be very short—and this is just as critical for the man as for the computer.

Here comes the issue of sizing up the central memory. The terminal must allow data entry, even if at that moment its microprocessor is occupied, for instance, in data exchange with other terminals. This concurrent type operation is one of the factors to be considered in choosing terminals. Available controls are another factor of choice. Errors because of the operator, the terminal itself, or the line must be taken care of.

Transmission must be controlled by the terminal, without the operator needing to worry. This is a function included in a procedural block which is part of the terminal's structure. Design must, in general, observe prerequisites which at times complement one another, while in other cases, call for contradictory characteristics. Such prerequisites include:

• direct communication capability with the central and distributed computer resources

• use of an easy to learn programming language

• simple overall structure, and devices accessible to man

• accuracy of data collection and transmission; authentication, authorization

• possibility of controlling the communication discipline

• obtaining the reply in a form the operator and even the end user can easily understand

A comprehensive classification of the types of application will invariably influence the number and kind of desired characteristics. The specific work to be done might suggest, for example, the wisdom of using softcopy rather than a hardcopy (or "blind") terminal.

The code(s) a given terminal is designed to use, the conversational methods, the protocols employed, the type of connection it can support, the speed of transmission are influenced by and, in turn, do affect the sophistication of the projected applications.

Other issues are the terminal's compactness (footprint), desktop and desk-bottom possibilities, modularity, and so on. When the volume of data to be handled is the prime factor in choosing a suitable terminal, the elements to bear in mind are: the presence or absence of memory and of storage devices, the ease of access and of operation, the printing speed, the ability to serve online/offline applications, and recovery capabilities.

Recovery is the possibility of functioning again after a falldown, without any loss of data or permanently interrupted connections. This can be assured through software (which, in turn, means a programmable terminal); it also requires online storage devices (such as floppy discs or cassettes). We will proceed further with this subject when we talk of line transmission but first we must define the issue of codes and coding.

17.5   THE USE OF CODES

Codes are used in many places in daily life, very often without our knowing it. This is the case with traffic lights, but also with the key one uses to enter his own home.

Music script is a code, the blind use Braille for reading, and computers use programs which are also written in a coded language. Codes are employed to identify chemical reactions. And the reader is well aware of the code numbers he employs for automatic self-dialing and in the mailing system, in the form of postal codes.

Etymologically, the word *code* has more than one meaning:

1. It is an orderly group of digits or signals with a one-to-one correspondence between two pieces of information.

For every piece of detailed data, there exists one or more codes—for instance, your friend George Brown living at 143 Lexington Avenue, NYC. An example is his telephone number (212) 997-4343. Another is his Social Security number.

2. It is a signal or a set of signals representing letters, numerals, special characters (such as %, /, &) used in sending messages. The use of flags is yet another example.

The telegraph service uses five-letter code words for the abbreviation of sentences, repeat questions, reannouncements, and the like. In any and every case, a description (function) corresponds to a signal (argument). Mathematically, we can also talk of a picture and of a representative illustration.

This description identifies the content of the code. A connection between argument and function can be found in comprehensive tables, or the function can be created through the use of key rules governing the coding of words and of messages. In chemistry, the elements are represented by signals which consist of one or two letters, for instance:

- O (Oxygen); H (Hydrogen); Ag (Argentum . . . Silver); Ca (Calcium); etc.

The description of chemical reactions is formed by these signs, for example: H_2O . . . Water. Vehicle nationality identification is composed of a one-to-three grouping of letter: USA . . . United States of America; UK . . . United Kingdom; GR . . . Greece; D . . . Germany; F . . . France; I . . . Italy; etc. But there are also other definitions given to the word "code":

3. The identification of a collection of items (for instance, in statistics a population) by means of abbreviation following mathematical rules: the mean, the range, and the like.

In this case an extended form is reduced by means of coding in a representative, mathematical, or pictorial form, following given rules.

4. The writing of a computer program in a form understandable by the machine (machine language) to guide the work of the computer—which at machine level language is also a binary code.

Signs should not be confused with codes. They are the basic building block from which bigger units, such as words can be formed. Examples of signs are: letters of the alphabet: A, B, C . . . or a, b, c—differentiating between capital (upper) and small (lower case) letters; decimal numbers: 0, 1, 2 . . . 8, 9; punctuation marks: . , ; ! ? ; instruction signs as: carriage return, new line, hyphen; hieroglyphics (picture writing); colors: e.g. red, orange, yellow, green, blue, white, black, and so on, when used to identify an entity. In traffic lights, for example green means "GO"; red "STOP".

The Alphabet (Character Set) which we use in written communications is an orderly collection of signs. The character set includes:

- letters of the alphabet
- decimal numbers
- punctuation marks
- instruction signs

The single letter: A, B, C . . . 1, 2 . . . is an element of the character set. The letter will be used, by means of procedural rules established by the language, to create words. The words are the building blocks of sentences and the sentences of paragraphs.

This is a brick and mortar approach, the mortar being syntax. That's the easiest way to explain how in data communication and data processing we are constructing the *information elements*. As we will see, the identification of information elements is an important step of the systems effort.

With computers, a fundamental set is the binary code. It presents the user with two alternatives: 0, 1; yes, no; black, white. This condition with two states finds counterparts in electronic equipment:

	State "O"	State "1"
through a wire can pass:	current	no current
a switch is:	open	closed
a transistor is:	conducting	non-conducting

Designers put to profit this binary feature of materials and components to construct information machines. But since the most common numerical system, the decimal, has 10 states (0 to 9) and not two, it is necessary to construct a code which represents these ten different states in a binary form.

There are many possibilities for coding the 10 decimal numbers. One example is the so-called 8-4-2-1 code, in which use is made of the *positional notation*, the way we know it with the decimal system. (See Figure 17.1) The positional notation and the concept of zero (giving to nothing the meaning of something), are the two pillars on which rests the mathematical (arithmetic, algebraic, algorithmic) advances of all time.

The 8421 code (with odd-parity).

Character	Significance 8	4	2	1	P
0	0	0	0	0	I
1	0	0	0	I	0
2	0	0	I	0	0
3	0	0	I	I	I
4	0	I	0	0	0
5	0	I	0	I	I
6	0	I	I	0	I
7	0	I	I	I	0
8	I	0	0	0	0
9	I	0	0	I	I
A = 10	I	0	I	0	I
B = 11	I	0	I	I	0
C = 12	I	I	0	0	I
D = 13	I	I	0	I	0
E = 14	I	I	I	0	0
F = 15	I	I	I	I	I

Figure 17.1

The decimal number of 2053 is written in terms of the positional notation of each digit (thousands, hundreds, tens, units):

$$2053 = 2 \times 10^3 + 0 \times 10^2 + 5 \times 10^1 + 3 \times 10^0$$
$$= 2 \times 1000 + 0 \times 100 + 5 \times 10 + 3 \times 1$$

Quite the same procedure is followed with the positional notation in the binary system:

$$1001 = 1 \times 2 + 0 \times 2^2 + 0 \times 2^1 + 1 \times 2^0$$
$$= 1 \times 8 + 0 \times 4 + 0 \times 2 + 1 \times 1$$
$$= 9 \text{ (in the decimal system)}$$

Hence, four binary digits (bits) are enough to code the 0 . . . 9 decimal digits. Since 16 combinations are possible with four bits, four combinations (from coding the decimal 10 to the decimal 15) will not be used with a computer circuitry tuned to decimal representation. The system using binary digits to code decimal digits is known as BCD (Binary Code Decimal), and is extensively employed in data processing and data communications.

The BCD system can be extended to code the letters of the alphabet. We said that four bits ($2^4 = 16$) serve as a code for the 10 decimal digits and leave six combinations unused. This is not enough to code the alphabet.

Five bits offer 32 possible combinations ($2^5 = 32$), hence, sufficient for the coding of all letters of the Latin alphabet. But the 32 different combinations will not be enough if we add the ten digits. We must, furthermore, provide for the coding of the needed operational signs, for instance, punctuation. Depending on the number of signs we wish to code, even 16 binary bits may not be enough.

Let's examine what we get with seven binary digits ($2^7 = 128$). The code range will be complete with 128 different characters to cover the 26 letters of the alphabet, the ten decimal digits, to reserve some positions for foreign alphabets (Hebrew, Greek, Russian, etc.) and to include some punctuation and other signs. But the available possibilities suggest that punctuation will be limited, hence, the interest in the ASCII 8-bit code.

The reference to the grouping together of bits is important. The storage (write) and the retrieval (read) to and from memory is always done in "a block of bits." This block of bits, called a *word*, may contain 8, 16, 24, 32, 36 or 72 bits. A *byte* is a term the reader will encounter often in literature. It is the standard packaging of eight bits.

An 8-bit code can have seven bits as message carriers and 1 for parity. Alternatively, the eight bits in a byte can all be used as message carriers ($2^8 = 259$) and a 9th bit is added for parity. Typically, this bit will be added where the greater possibility for error exists: in data transmission and in peripheral storage media (such as magnetic tape) rather than when data is stored in the central memory of the computer.

The reader should nevertheless know that, unfortunately, the introduction and use of codes is not standardized. This complicates the exchange of computer support media such as magnetic tapes, magnetic discs, and floppies. Some manufacturers have developed their own codes. Even machines coming out from the same manufacturer, with similar kinds of supports, utilize different coding procedures. This handicaps compatibility.

17.6   FACTS BEHIND STANDARDIZATION

To better appreciate the need for standardization and the confusion because of the lack of it, it helps to take a historical look at the development of this basic data processing tool: character codes. The encoding of character sets, used for data handling purposes, since the time of Morse code for telegraphy, has taken many forms. A set of characters did not become permanent and interchangeable until the advent of punched tape and punched cards, respectively representing the continuous and discrete media.

Punched cards were originally used for accounting and statistical applications at first only requiring a 12-character set—10 digits and two signs. From the beginning punched tape was used for message communication, and the 5-channel (track) tape offered 60 characters: $2^5 = 32$ possible characters but further encoding made up the difference. Thirty (30) encodings had double meaning following Figure Shift and Letter Shift, and so on.

This code was because of Baudot, and the assignment of codes to characters was made on the basis of letter usage frequency, to conserve electrical energy. Similarly with punched cards, two alphabetic zones (the 11th and the 12th, or A and B rows) were added to make feasible alphanumeric coding. (Since the use of information machines at the earliest time was computational, the punched card set did not receive much impetus for enlargement until applications started with business-oriented problems.)

With tapes, cards, cassettes, and floppy discs, standardization was left behind. Furthermore, starting in 1954, the need for mathematical and arithmetic symbols to take encodings as that of business symbols led to much confusion in installations that did both types of work.

The Baudot Code, incidentally, was not without its mutations, although the CCITT* Working Alphabet Nr. 2 was used for interchange throughout much of the world. Even so, the binary positional notation was perturbed by different assignments of bits to the tracks.

By the late 1950s it became apparent that code expansion had grown uncontrolled. There were more than 20 variations for the punched card code, and more than 60 different internal interpretations in computers. Some early pressures for code expansion came from programming languages. Univac I, the first computer for business usage, had a 51-character set on its printer. The IBM set had a mechanical keypunch limitation of 48 characters, mostly business symbols.

Pressures arose for unique encodings. In 1956 it was decided that the IBM Stretch computer would have a 64-bit word, and that the characters would be represented by eight bits. This decision had much influence on subsequent computer architecture—and on the coding schemes as well.

Slowly, the real need for standardization became clearer. By 1959, several independent standardization projects began, yet this year there were nine different

---

*Consultative Committee International for Telegraph and Telephone of the International Telecommunications Union (ITU).

internal codes existing in IBM equipment alone. Interchange on media was a problem throughout the world. Associations got into the act, not only related to computers, but also to business machines at large. In the U.S., the Office Equipment Manufacturers Institute was revitalized and converted to the Business Equipment Manufacturers Association, and BEMA acted as a sponsor for the American standards effort in data processing for ASA (American Standard Association, now ANSI).

In ISO (International Standardization Organization) it was Sweden that recommended in late 1959, activities in data processing standards. This led to a Round Table Conference in Geneva, on May 16, 1961, which in turn led to the formation of the ISO Computers and Information Processing task force. And the European Computer Manufacturers Association (ECMA) also got into the act.

The ISO-alphabet

	OO	OI	IO	II
.. O O O O		O	NUL	P
.. O O O I	f	1	A	Q
.. O O I O	f	2	B	R
.. O O I I	f	3	C	S
.. O I O O	f	4	D	T
.. O I O I	f	5	E	U
.. O I I O	f	6	F	V
.. O I I I	f	7	G	W
.. I O O O	(	8	H	X
.. I O O I	)	9	I	Y
.. I O I O	*	:	J	Z
.. I O I I	÷	;	K	n
.. I I O O	,	n	L	n
.. I I O I	—	n	M	n
.. I I I O	.	n	N	ESC
.. I I I I	/	´	O	DEL

ESC (= Escape)
DEL (= Delete)
n. . .Reserve for "National" signs
f. . .Formating sign

Figure 17.2

By the mid-1960s, there was some progress toward standard codes, but not necessarily to a single standard code. As invested interests did not allow the adoption of a unique solution, equipment was designed to handle both the extended BCD Code (for upward compatibility of much former equipment) and the eventual ASCII. These two codes represented the foremost in standardization for over 15 years.

The American Standards Association has published the ASCII coded character set to be used for the general interchange of information among information processing systems, communications systems and associated equipment. (See Figure 17.3) An ASCII 7-bit code looks like this in hardcopy:

	Symbol	b_7	b_6	b_5	b_4	b_3	b_2	b_1
(Start of message)	SOM	0	0	0	0	0	0	1
	(F	1	0	0	0	1	1	0
Character Code for	(A	1	0	0	0	0	0	1
	(S	1	0	1	0	0	1	1
(End of message)	EOM	0	0	0	0	0	1	1

American Standard Code for Information Interchange

b_7	b_6	b_5	b_4 b_3 b_2 b_1	O O O	O O I	O I O	O I I	I O O	I O I	I I O	I I I
			0 0 0 0	NULL	DC_0	+o	O	O	P		
			0 0 0 I	SOM	DC_1	!	1	A	Q		
			0 0 I 0	EOA	DC_2	"	2	B	R		U
			0 0 I I	EOM	DC_3	#	3	C	S		N
			0 I 0 0	EOT	DC_4 (STOP)	$	4	D	T	U	A S
			0 I 0 I	WRU	ERR	%	5	E	U	N	S
			0 I I 0	RU	SYNC	&	6	F	V	A S	I G
			0 I I I	BELL	LEM	, (APOS)	7	G	W	S	N
			I 0 0 0	FE_0	S_0	(	8	H	X	I	E
			I 0 0 I	HT SK	S_1	)	9	I	Y	G	D
			I 0 I 0	LF	S_2	*	:	J	Z	N E	
			I 0 I I	V_{tab}	S_3	+	;	K	[	D	
			I I 0 0	FF	S_4	, (KOMMA)	<	L	\		ACK
			I I 0 I	CR	S_5	–	=	M	]		①
			I I I 0	SO	S_6	.	>	N			ESC
			I I I I	S	S_7	/	?	O			DEL

Figure 17.3

Procedural descriptions (also worked out by ANSI) indicate the means of implementing this standard. The 1982/83 development of the North American Presentation Protocol Syntax (NA/PLPS) has been a major step forward. It defines graphic sets and includes an ASCII 8-bit code in realization of the fact that computers and communications are now predominantly using 8-bit characters on 9-channel magnetic support media.

(Storage on magnetic tape often follows the 9-bit code which results by adding one parity bit at the end of a string of eight bits. These nine bits are stored as a self-contained unit. While reading a magnetic tape, the computer automatically checks for possible writing or reading errors.)

The standardization of codes is one thing; that of protocol layers another. Though the subject of protocols is treated in the next chapter, to provide a well-rounded discussion on standardization we must add an introduction to the layered concept at this point.

For data processing and data communications purposes, ISO has introduced a general structure known as *Open Systems Interconnection* (ISO/OSI). It provides architectural concepts from which a reference model can be derived by making specific choices for the layers and their contents. From higher to lower, the seven layers are:

- Application (User Level)
- Presentation Control
- Session Control
- Transport
- Networking
- Data Link, and
- Physical

It may be difficult to prove that any particular layering selected is the best possible solution, but there are general principles which can be applied to the question of where a boundary should be placed and how many boundaries should be used. Among the principles to be considered is the avoidance of creating so many layers as to make difficult the system engineering task describing and integrating them—as well as to set the boundaries at points where the services description can be small and the number of interactions across the boundary are minimized.

ISO/OSI aims to create separate layers to handle functions which are manifestly different in the process performed or the technology involved, collect similar functions into the same layer, and select boundaries at a point which past experience has demonstrated to be successful. It is creating layers of easily localized functions so that each could be totally redesigned and its protocols changed in a major way to take advantage of new advances in architectural, hardware, or software technology without upsetting the services and interfaces with the adjacent layers.

Interfaces are created where it may be useful at some point in time to have a layer standardized, insert a new layer when there is a need for a different level of abstraction in the handling of data, (morphology, syntax, semantics), and enable changes of functions or protocols within a layer without affecting the other layers.

ISO/OSI permits usage of a realistic variety of physical media for interconnection with different standards (e.g. V.24, V.25, X.21. . .) fitting within the *Physical Layer* as the lowest layer in the architecture.

Physical communications media, such as telephone lines, require logical techniques to be used in order to transmit data between systems despite a relatively high error rate not acceptable for the great majority of applications. These techniques are used in data link control procedures, leading to identification of a *Data Link Layer*.

Furthermore, some systems will act as final destination of data; others only as intermediate nodes, forwarding data to other systems. Normalization leads to a *Network Layer*. Network oriented protocols such as routing will be grouped in this layer, providing a connection path (network-connection) between a pair of transport entities.

As the network layer uses sublayering, the common functionality in terms of facilities is employed to operate the Transport Layer. Control of data transportation from source end system to destination end system is the next function to be performed in order to provide the totality of the transport service.

The upper layer in the communications part of the architecture is the *Transport Layer*, sitting on top of the Network Layer. It relieves higher layer entities from any concern with the transportation of data between them.

The transport layer is followed by higher-up levels: the *Session* and *Presentation Control*. CEPT and PLP norms address themselves to the sixth layer of ISO/OSI: Presentation Control. The object is to:

1. Follow-up on session establishing, maintaining and terminating (toward the lower layers).

2. Request the creation of a process (toward the upper layers).

3. Notify processes upon receipt of data—making data more understandable through segmenting, blocking, and so on.

4. Assure message management beyond the level of buffering and error controlling.

5. If virtual terminals are used, adapt requests to the specific machinery at that location.

6. If programs use local names, translate them to a common reference, and vice versa.

7. Finally, provide for information enrichment, encrypting/decrypting, compacting/decompacting.

These are logical functions whose performance aims to assure an effective end-to-end communication between data terminating equipment. We will look in further detail into this issue when we talk of protocols.

Chapter 18

THE USE OF PROTOCOLS

18.1 INTRODUCTION

Protocols are formal conventions guiding the transmission and handling of information. They are implemented in the computers or other units connected to the network to help control data transfer from one device to another using the facilities which are offered by the hardware elements.

Protocols function at various levels:

- from low-level byte and packet transport
- to high-level database access in the applications layer

As such, protocols are an integral part of all types of computers and communications networks. The formalisms supported by protocols help regulate, among other issues, flow control and error detection.

Protocol formality and standardization are most important. Protocols must be compatible for each layer of the ISO/OSI model to communicate with its corresponding layer residing in another equipment. This can be achieved:

1. by adopting the same protocol at each corresponding level
2. by performing appropriate protocol transformation at a gateway

Care must be taken to assure that needed capabilities are not sacrificed for the sake of ease of the network interconnection, and that no errors are made in translation.

We said that the protocols can be of high or low level. The term "low-level protocol" identifies formalisms close to the line discipline, hence, used to transport groups of bits (or bytes) through the network. A low-level protocol is not aware of the meaning of the bits being transported. A high-level applications protocol knows better about them as it uses the bits to communicate about remote actions.

At the low level, high performance is achieved through hardware technology, enabling the simplification of protocols. They are designed to take advantage of and preserve the special capabilities of the layer to which they belong, so that they can be employed by higher level protocols to gain full advantage of the facilities being supported.

This reference exemplifies how valuable is *layering* as a design issue. It is a technique to assure logical independence of network modules, so that different communications subnetworks, interprocess communications, and data management functions can be altered with minimum disruption.

Layering had also its cost. If the layers are multiplied they may result in an explosion of control fields and interface requirements. Such control fields belong to the various layers and bring a drastic increase in communications overhead.

Let's conclude this discussion by underlining the importance of protocols in the future of computers and communications. Neither hardware nor software is central to the study of man-information communication channels, though both are certainly necessary for implementation. The most fundamental component is the *communications protocol*. (See Figure 18.1.)

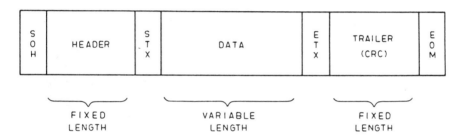

Figure 18.1 Structure of a Communications Protocol.

18.2 PROTOCOL FUNCTIONS

We said that in an attempt to standardize network architecture, the International Standards Organization (ISO) has defined a model of a network that consists of several layers (OSI).

The bottom layer is physical and defines the electrical interface and includes such parameters as the transmission medium, data rate, and mechanical interconnection specifications. Here the protocol is expressed in hardware, while for all other layers it is in software.

We also briefly stated in the preceding chapter that immediately above the physical link lies the data link. The protocol is implemented via a hardware/software combination. The data link layer supplies services such as message framing, error detection, and addressing.

Data link functions can be handled through either of three types of protocols:

1. *Start/Stop*, or asynchronous, is the eldest type.

Transmission is done character by character preceded by a start bit and followed by a stop bit. This protocol typically uses low speed lines, calls for low cost implementation at the device and modem levels, but is expensive in terms of communications costs because of its relative inefficiency.

Still, the largest part of installed terminals work start/stop. Let's also add that for non-intelligent terminals this is the only option, as any other data link protocol implies some programming capability at the data link level.

2. *Bisynchronous* (BSC). Introduced in the late 1960s, this is the first data link protocol which made use of a packet structure.

A packet features a header, which says "where from," "where to," and other control information; a main body carrying the data to be transmitted; and a trailer. The function of the trailer is error control.

With BSC, error control functionality covers only the data part, not the header. Furthermore, BSC is in a way associated with the message which it transmits. It does not necessarily optimize network capacity. This is done by another synchronous protocol in the:

3. *Packet Switching* discipline. At the data link level, the protocol advanced by ISO is known as High Level Data Link Control (HDLC), but there are also other offerings such as Synchronous Data Link Control (SDLC) by IBM.

Like BSC, a packet structure has a header and a trailer, the latter covering in terms of error protection both the data and the header. Furthermore, a packet structure is like a container. A message may have to be divided into more than one packet, while a packet may include (in its data part) more than one message.

Most important is the fact that a device using one protocol cannot ordinarily communicate with a device using another protocol. Each protocol differs in the specific formatting of its own *handshake*, and that difference precludes communication. To permit exchange of information, a protocol converter is required.

The converter is a black box that alters the format of one device's protocol to conform to the protocol format of the other device. For instance, it converts an asynchronous protocol into a bisynchronous format, thereby permitting communications between devices that otherwise could not exchange data.

Protocol converters, known as *gateways*, function primarily as translators, allowing one terminal to understand and respond to the message initiation of another terminal. This is essentially accomplished through software functionality, and helps establish an extensive communications link.

The discussion on protocol converters brings up the issue of packet assembly/ disassembly and associated support for unintelligent terminals. Such terminals are:

- capable of sending only one character at a time
- have no logical processing power with which to implement protocols

In Arpanet this support process is called a minihost. It is handled in the terminal interface message processor (TIP: kind of terminal concentrator), which also serves as a network switch.

The network, the user, or a third party, can also provide the support function through a separate terminal support device, such as a frontend or terminal controller that connects to the "TIP," acting only as a switch (in this case, "TIP" is a node).

The *Packet Assembler*, as the realization of this support function is called, addresses itself to the standard packet switching protocol in order to communicate through the network. In addition, another protocol must deal with the issues of terminal support between the packet assembler and the destination host.

The host may need to be informed of, or to be able to set, certain terminal parameters such as: duplex, code, speed, and packet termination conditions. Other parameters may result from specific implementation requirements.

The last references have brought under perspective the networking protocols. Above the data link layer is the networking layer, implemented in software. Tasks relegated to this layer would be routing and the implementation of virtual circuit (VC) or datagram—respectively dividing into two sublayers.

Among the responsibilities of networking are the division of a message into the appropriate packets, the retransmission of a packet, and the building of a whole message from a group of packets sent separately.

With a virtual circuit implementation, the last function is executed at the network node level. With a datagram, responsible for it is the data terminating equipment.

Networking protocols can be mapped into firmware and microprocessor controlled devices. Indeed, the effects of standards on dedicated chips are clearly illustrated in communication networks. The acceptance of X.25 (a virtual discipline) as a universal standard for public data communications networks paved the way for the dedicated X.25 protocol controller. (One of the commercially offered devices, Western Digital's 2501, carries three microprocessors along with ROM and programmable logic arrays.)

The higher of the communications levels is the Transport Layer. It performs three basic functions.

First, it segments the information elements (data units) it has received from the session layer (residing at the data terminating equipment) into the appropriate forms it passes on to the network layer. Second, it maps the session connection which has been established into a transport connection by:

• Working end-to-end

• Providing a reliable delivery service

• Assuring sequence control, flow control, error detection and recovery at the datacomm end

Third, with X.25, it allows the layer below to use packet switching. With alternative line disciplines, a similar statement can be made for circuit switching, message switching, or whichever other approach is supported.

18.3   OUTLINING SUPPORTED SERVICES

Each layer at a typical communications node includes one or more modules that provide the desired network services at that node. Many networks have identical modules in each layer of each node; but this is not a universal practice. Other networks contain nodes with special modules.

In the general case, the supported services in a communications network rest on protocol interpretation and implementation. The description and verification of protocols is an important subject in any network. Protocols must be unambiguous and formal methods should be available for their handling. The implementation of protocols must be carried out automatically.

As a set of rules to control interactions between two or more entities, a protocol is subject to a formal description. Like any finely tuned mechanism, it should feature a finite set of:

1. internal states

2. input signals from the network management and the execution environment

3. input signals from both the higher and the lower layers

4. output signals to the same entities (network management, executable environment, higher and lower layer)

The formal description and representation of protocols is a prerequisite to their utilization. For manipulation purposes, a protocol compiler assists in acting as a sort of translator which generates the object from the source program.

Protocols are required for synchronization purposes:

• *At the bit level*, so that the receiver knows when a character starts and ends, so that it can be sampled;

• *At the character level*, so that the receiver can determine which bits belong to a character;

• *At the message level*, so that the receiver can recognize the special character sequence which delineates messages.

Typical protocols include the blocking of transmission into messages employing start-of-text and end-of-text, (STX/ETX) or other similar markets and a positive/negative acknowledgment (ACK/NAK). Table 18.1 outlines the most important standard symbols in data communications (ASCII). Figure 18.1 presents the standard format of a message.

ACK is the positive acknowledgment. NAK is negative acknowledgment. SOH is start of header and prefixes control information at the beginning of each message. STX indicates the start of the message and implies the end of the header, if there was one.

ETX marks the end of the message, and is the last character in the message, unless we are using a block parity check. EOM is the end of message. EOT is

TABLE 18.1 Standard symbols in data communications

ACK   —Acknowledgment
BEL   —Audible Signal
DCo   —Device Control Reserved for Data Link Escape
DEL   —Delete/Idle
DLE   —Data Link Escape
ENQ   —Enquiry
EOA   —End of Address
EOM   —End of Message
EOT   —End of Transmission
ERR   —Error
ESC   —Escape
ETB   —End of Transmission Block
ETX   —End of Text
FE   —Format Effector
LEM   —Local End of Media
NAK   —Negative Acknowledgment
NUL   —Null/Idle
SOH   —Start of Header
SOM   —Start of Message
STX   —Start of Text
SYN   —Synchronous Idle
DEOT   —Mandatory Disconnect, used to force the disconnection of a switched connection. It is
       the sequence DLE, EOT.
ACK-N —Acknowledgement N, used to provide numbered ACKs. ACK0 is DLE, 0 and ACK1 is
       DLE, 1.
BCC   —Block check character, used to provide LRC checking.

BCC follows ETB or ETX and sums over the entire message, exclusively or the starting SOH or STX, but including the ETX.

the end of transmission, but can also be used to reset the line to an agreed starting point in the protocol.

ENQ is used to solicit a response from the other station; the exact meaning will depend upon the context. The sequence is always begun by sending ENQ to determine if the other station is ready to receive a message. If it is, it will reply ACK. If not, it will send NAK.

Some characters are used to form messages; others to control the transmission of messages. On receipt of ACK, the sender station can go ahead and transmit the message. The receiving station will check each incoming character for parity.

If the message is correct, it will send back an ACK and the sender can transmit another message if it wishes. A NAK to a message will cause the originating terminal to retransmit.

When the sender has sent all the messages for a particular exchange it will terminate with an EOT. This character may also be used by the receiving terminal to terminate the exchange if local requirements (or problems) warrant it.

A protocol, for instance SDLC, will define the composition of a message: the heading and trailing sections, the controls, and the message proper. It will identify the control information (contained in that portion of the message called the heading), and generally delineate the message. Some typical heading information is:

- identification of the originating station
- identification of the sending and receiving device or process
- the priority of the message
- the security class of the message
- routing information concerning the distribution of the message once it has reached its destination
- message processing information concerning its status as data or control

We said that packet switching protocols support software over checking procedures for data integrity. A two-byte cyclic redundancy check is employed for error detection. The bits are often generated through hardware in host computers, and their calculation is usually protocol dependent.

With a software approach, a general purposes hardware interface can handle all synchronous mode data communications. Modifications will accommodate protocol dependent characteristics. An n-bit data block to be transmitted is treated as a binary polynominal of the form

$$F(\chi) = b_n + b_{n-1} \chi + b_{n-2} \chi^2 + \ldots + b_1 \chi^{n-1} + b_0 \chi^n$$

So much for the basic communications control procedures used for establishing, maintaining, controlling, and terminating a communications link between two processes, two equipments, or two persons communicating through these equipments. In large, complex networks it is desirable to build upon these basic procedures and form higher level protocols, to control the flow of messages between processes, distributed among the many computers and devices which may be interconnected by the network.

As a protocol is a set of conventions between communicating processes, to make implementation and usage more convenient higher level protocols use lower level protocols in a layered fashion. As an example, higher level protocols include:

1. *the initial connection protocol*, which can be used by all processes to initiate a communication link with remote processes;

2. *the remote job entry protocol*, for communicating between devices and processes of the remote batch variety;

3. *the file transfer protocol*, for the transmittal of large files from one computer to another;

4. *other protocols defining a virtual terminal*, thus permitting all terminals on the network to provide a similar interface to processes in any host.

These protocols allow the use of any terminal supported by the network, with any host system on the network, even if that host could not support the particular terminal through its standard terminal controller.

18.4 PROCEDURES FOR PACKET SWITCHING

Packet switching is the best technology available today to accommodate a range of transmission demands. It offers potentially better network utilization than circuit switching. However, it also implies the implementation of information gathering and routing algorithms requiring intelligence at the node and/or terminal level.

Circuit switching is the elder approach implemented with voice grade lines originally used with classical electromechanical switching centers. A wire circuit or radio link connects the parties. Either point to point transmission or multidrop discipline can be implemented, the latter by means of Polling protocols (polling-selecting, hub polling).

As contrasted to circuit switching, message switching is a more recent development dating to the post-World War II years. Telex/TWX has been its first implementation (1940s); Swift, the global network for international financial transactions, is another example (1970s).

With message switching, a fairly intelligent message switch is necessary; storage facilities must be supported at the node level for store and forward; no direct wire connection is necessary; broadcasting and multiaddress of messages is feasible; and the network can perform speed and code conversion. Message switching and its facilities are applications-oriented.

Packet switching improves upon message switching both resulting from the protocols and technology being used and also, most importantly, because the network is of a generalized type.

• The packet switching network is not concerned with the applications content of the transported data.

• The route is established dynamically for each packet.

• Contention for network resources is significantly reduced, while dependability on message delivery is increased.

A greater buffer space may be needed with this approach, but today incorporating additional buffers is considerably less expensive than adding network paths. Larger buffers at the nodes and data terminating equipment impact on the implementation of the network, but are not overly costly to effect.

Within the overall packet switching perspective, each packet has a source and destination, both of which are identified in the packet's header. A packet placed on the bus by the sending node propagates to the destination node. Any interconnecting link can carry this packet but normally only the intended destination station matching its address in the packet's header will copy the packet. The way these services will be supported at the packet header varies from one implementation to another. With X.25, the first two bytes of information in the packet header, or more accurately, the first four bits, and the next twelve bits, are used to identify the destination of the packet.

- This is the logical channel number (LCN) or logical channel identifier.

- The four bit field marked ''ID'' can have only two values, either 0001 or 1001.

- 0001 signifies that the packet is being transmitted to an individual terminal or computer directly.

- 1001 is used for sending a transmission to a device via a cluster controller.

The issue of the addressing protocol is important. Datagram must contain the entire address. Virtual Circuit involves identification (ID) and logical channel number (LCN).

The addressing protocol is the responsibility of the network administrator or the terminal handler. This is done at the node, or at the terminal itself.

- *If* after the node, the discipline is polling/select, then addressing is a node responsibility (or the concentrator's).

- *If* the point-to-point terminal supports X.25, then each DTE must have the needed network administration and terminal handler software.

Every data terminating equipment supporting X.25 must handle the addressing requirements. For an X.25 device, the minimum central memory is 16 KB and growing. More software in memory is necessary to handle network addressing, data links, and the logical part of electrical interfaces (Figure 18.2).

The node responsibility includes:

1. Recognizing the presence of a packet.

2. Examining the service information, including the header.

3. Identifying eventual errors in the transmission.

4. Directing the packet to the right destination.

5. Assuring network journaling.

This calls for intelligent nodes. The node must also handle the link connection and the needed protocols to that end. There are two fundamental types of virtual circuits, ''permanent'' and ''switched.''

A *permanent virtual circuit*, linking—for instance, terminal to computer—is where a logical channel number is assigned by agreement at the time the service is initialized. This is not a permanent physical circuit, but the priority it carries is high and, though logical, it acts *as if* it were physical.

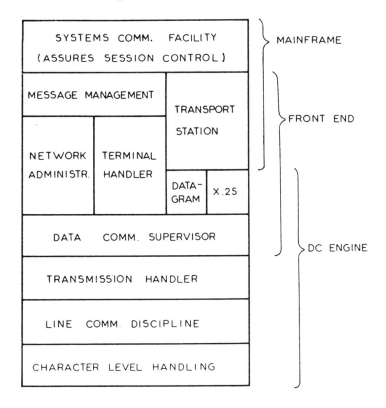

Figure 18.2 Layers in a Communications Discipline.

A *switched virtual circuit* is under control of the terminal and of the network (dynamic solution). Hence, a mechanism is needed and this is the call request packet.

The call request packet allows the user to place a call to a specific terminal. Each such attachment is assigned a number, like in the telephone network. The call request packet assigns a currently available logical channel number (LCN). This number will be active for the duration of the call. Then it will be released.

LCN is a brief number, as compared to the generally much longer DTE attachment number. But at the completion of the call, LCN is released for reuse. With 12 bits available for LCN, there are up to 4,096 virtual circuits which can be supported simultaneously. The word "simultaneously" means over a single physical connection to the network.

The control field within the packet header plays a role similar to that of the control field in the link header. It indicates whether:

1. This packet is a call set up request or disconnect.

2. This is a data transfer.

3. An interrupt of the devices at the other end is required, or

4. The flow of data should be temporarily stopped.

It is also used in error detection and correction and in the handling of exception conditions.

18.5   ROUTING

Messages are transmitted in a given form, through an established path. This path is the *Route*. Routing is the process of creating the route.

• A *fixed route* involves one, and only one, path between two stations in a network.

• A variable of *dynamic route* presupposes the capability of switching.

Point-to-point and multipoint connections are examples of fixed routing. They are disciplines used with classical circuit switching and do *not* involve the services offered by packet switching.

Dynamic routing coinvolves tables in the memory of the node/switch for the optimal use of resources, congestion and flow control, delay analysis, and so on. Routing procedures must respond to two basic questions:

1. *What's to be done at the node level?* Traffic requirements, "reachability" matrix, etc.

2. *Efficiency criteria to be observed:* reliability, availability, accuracy, error detection and correction, and adaptability.

Adaptive routing provides optimization in terms of delays, assurance of shared resources at node and link levels, better throughput, and lower overall cost for same service level. Table 18.2 presents input/output data for routing needed at a node switch.

Another reference is pipelining. Networks must allow multiple message transit between source and destination. This permits higher reliability and good through-

TABLE 18.2 Input/output data for routing

The Matrix at Node Switch may take the form	
Routing Input Data	Routing Output Data
1. *Adjacency Matrix* Existence of lines between nodes	2. *Reachability Matrix* Existence of paths between nodes
3. *Traffic Requirements* Binding of traffic to destination nodes	4. *Traffic Assignment* Total path perspectives

put. At the same time, throughput considerations must be balanced with distance: source to destination, line speed, and size of packet.

Other procedures must also be observed in establishing a routing discipline. Will its basic characteristics be deterministic (fixed) or stochastic? Given that each node probes the communications system, will it or will it not share information with the other nodes? What about centralized control?

Heuristic type algorithms tend to be favored in network design. They provide more flexible approaches and a better utilization of resources. But they also require a greater intelligence at nodes switches.

In a properly distributed datacomm network, each node probing shares the information with the other node. This is generally true for the nodes with which it maintains a direct link—not network wide.

Packet switching networks are usually peer systems; but there are exceptions. A few packet switching networks opt for a pseudo-centralized control. With Tymnet, one master node calculates for the whole network the variable routing solution.

Routing must also be done in relation to cost evaluation. The cost elements include:

- nodal bandwidth

- nodal delay

- nodal storage

- line bandwidth

- line delay

Nodal bandwidth relates to a multiprocessing capability at the switching level. If routing and other functions are supported, *then* routing must have priority. *Nodal delay* is present because of the supported facilities: routing calculation, route matrix update, store and forward, error detection, and so on.

Nodal storage is necessary for the route matrix, the algorithms which will be used, and the I/O buffers for data transit. Here, the most important reference is store and forward.

Line bandwidth is the transport capacity supported by the connecting link(s), in kbps or Mbps. Its use is conditioned by the messages to be routed, choice of path, retransmission because of errors, length of the messages, and housekeeping information "en route."

Line delay increases linearly with routing message frequency. It increases quadratically with message length, but it decreases with line bandwidth.

18.6 FLOW CONTROL PROTOCOLS

We said that protocols exist for many reasons, transport control being one of them. To support interprocess communications, the transport control must reflect strategies which allow the input flow of messages to be limited in accordance with the reception possibilities of the destination entity.

A goal of flow control is to make the throughput closely related to the receiving capability of such entity. A mechanism similar to that used to detect message loss or duplication may be employed for flow control. Throughput depends on:

1. bandwidth
2. the nature of the attached devices
3. the processes running on them
4. their communications requirements
5. message size
6. retransmission timeout
7. buffer availability
8. allocation strategies
9. roundtrip delay

Several of these factors are interrelated. If the number of lost or damaged messages is not negligible and the transmission medium is not congested, retransmission timeout can be chosen to be short in order to perceptibly reduce the average transmission delay of messages. Since the sending process keeps a copy of each message transmitted but not yet acknowledged, the size of the buffer used by the sender limits the number of outstanding messages.

If the delivery of messages must be made in the same order as submitted, the receiver should keep the arriving messages in a buffer, as long as there is a message missing. This reduces the reception capability of the destination entity.

A roundtrip delay is the amount of time between the sending of a message and the return of the corresponding acknowledgment. Definition of the acknowledgment discipline is very important for flow control. Acknowledgments may be forwarded to the sender:

- systematically after each successfully received message
- following several received messages.

The latter discipline is used when systematic return of acknowledgments results in significant loading of the transmission medium.

When acknowledgement is returned only after receiving several messages in succession, the result is an increase in roundtrip delay associated with a successfully received message. Roundtrip delays may involve retransmission by the sender of messages which have already arrived at their destination.

Acknowledgment after every received message also has its drawbacks. Increase in traffic resulting from housekeeping operations is one example. A similar case can be made about the usage of resources at the destination entity.

Message size can have a direct effect on both the sending and receiving buffer area. In this sense it impacts flow control, and can significantly influence flow control performance. Large messages:

- Require a larger buffer space and greater amount of available bandwidth

- Cause more transfer errors—therefore more retransmissions

Short messages reduce propagation and roundtrip delays, and require a smaller buffer area, but the overhead is increased as header and control information is fixed length. With short messages, acknowledgments and other housekeeping chores are more numerous, for the same amount of data to be transmitted while switching time is fixed independently of message size.

Buffer availability is an important reference in most applications. Let's take as an example electronic mail. We must achieve at least one session for one mail. Moreover, a certain transfer delay may be requested by the user or is defaulted by the system.

During the imposed delay, it is possible to open one or several transport connections to transfer one mail, but this brings up a storage problem of the transported data units. Store and forward must necessarily take place between the two sessions layers connecting end users. If not, one session could only open a transport connection.

We have also spoken of roundtrip delays. The transmission delay time permits one to make best usage of the network's hardware resources, thus limiting the need to size the new network components according to the peak hours traffic. This last reference will be so much more valid if we care to monitor traffic statistics; query about connections to evaluate foreseeable traffic in certain elapsed delays; and verify whether a request to transfer data is compatible with the current and projected traffic load.

A different way of making this statement is that processes and their communications requirements must be examined in unison. A solid strategy for implementation is to keep text and data handling at the workstation (terminal) level, thus reducing communication to matters such as the updating of a common (local or central) database, the transfer of messages, and the necessary interprocess exchanges.

A similar reference can be made regarding the interconnection of attached devices in terms of their characteristics and the bandwidth they require. Mainframes and minicomputers pose different load characteristics than communicating intelligent terminals. In a local area network (LAN) environment, they call for broadband rather than baseband connections.

Hundreds of workstations on a network amount to a significantly greater load than a couple of dozen WS—though the nature of the work being done also weighs on the implementation algorithm, as the preceding paragraphs documented. Furthermore, bandwidth places constraints on the number of intransit messages.

If the carrier is loaded with intransit messages, any message retransmission will be made by reducing new data transmissions and, therefore, useful throughput. If the bandwidth is not entirely loaded with new incoming messages, any message retransmission resulting from loss or failure will be made without modifying the throughput in an appreciable manner.

In either case, control procedures must be implemented. For instance, detection of lost or damaged messages can be achieved by using sequential ordering of messages when submitted by the sender. The size of the sequence number space attached to each message will typically be finite. The receiving throughput, transmission bandwidth, and sequence number space impact on the protocol to be chosen (Figure 18.3).

ADDRESSING PROTOCOL

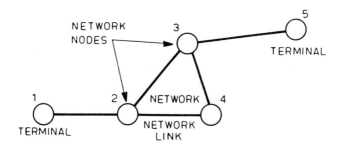

DATAGRAM MUST CONTAIN ENTIRE ADDRESS.

1, 2, 3, 5

VIRTUAL CIRCUIT INVOLVES ID, LCN (LOGICAL CHANNEL NUMBER)

ID
LCN

Figure 18.3

Chapter 19

KEY TOPICS IN DATA COMMUNICATIONS

19.1 INTRODUCTION

As explained since the beginning of this text, data communications is a fundamental, integral part of an online system. The transmission of data has prerequisites which can be performed in software or in both software and hardware. We will see which these prerequisites are. In principle, they can be answered from:

• The central computer, equipped with a *line control* faculty, which can be either an integral part of the computer (internal) or an attached unit (external) to the computer

• A special computer connected to the central computer and geared to line control problems (the frontend)

• The nodes switches of the network

• The regional computers, local computers, or intelligent terminals with a line controller—attached to this network

The user will rarely if ever see the difference between one technical solution and another. But he will feel the effects: delays, possible errors, code limits, difficulty to link to host. Line control units are an integral part of the computer (or terminal) the user acquires for data communications functions. Their hardware and (most particularly) their software establishes the range of supported services and the corresponding constraints.

The person sitting at a video unit, transcribing a message onto the keyboard, does not necessarily possess details of the technical components of the PC or terminal. Just the same, the user of a packaged product, for instance a radio or TV set, does not enter into the details of the units in its composition—though the technical nature of these units and the choice among alternatives is of great interest to a ham radio operator.

The driver of an auto does not necessarily know how the internal combustion engine works, or which choices exist for the brakes. But he or she does know of their existence except in a building fashion. For datacomm, this building block approach is shown in Figure 19.1.

Protocols, codes, information elements, line disciplines, operating systems, and network architectures are the logical components of the system. Computers

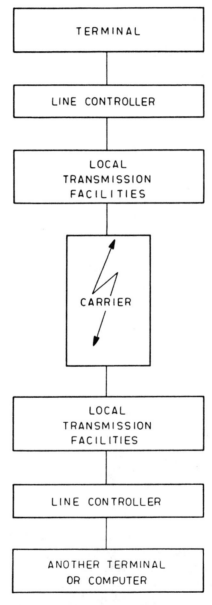

Figure 19.1 End-to-End Transmission.

(whether mainframes, minis or micros) terminals, and transmission lines are the physical components. Both physical and logical components are necessary to make up a system of data communications, able to connect two users end-to-end.

The users may be human operators, processes running on machines, or other computers/microcomputers/terminals. In either case, they must observe specific disciplines for a communication to be effective. The *data path* from the one computer (or terminal) user to the user computer (or terminal) user is a complex trail:

1. The terminal is connected to an adapter which performs a signal transformation, acceptable to the local transmission line. This adapter is the *modem*. More about this later, but let's state that it serves to connect the terminal (or computer) to the local network port. However, if the terminal is located close enough to the computer, say in less than a quarter mile, the adapter may not be necessary.

2. A set of transmission facilities interconnects network ports. At the computer site or, generally, the destination another network port is required along with the hardware connection between the computer and the port. Networks ports are interconnected through network lines.

3. The circuit proper, or carrier. Etymologically, a *carrier* is an electrical signal, chosen because of its ability to travel through the transmission medium being utilized. In the case of data communications, the medium is a telephone line. As related to telephony, the word "carrier" also has other meanings. Presence or absence of a carrier simply refers to presence or absence of transmitted energy. In a much broader sense, the word "carrier" is used by extension, as synonymous with the telephone company—the organization which assures that carrier, in the sense of an electrical signal, has the proper facilities it requires for functioning.

The obvious function of the carrier is to carry the data. This is done through the telephone wires (see the discussion on Simplex, Half Duplex, and Full Duplex). These telephone wires are often called a circuit. There are circuits that:

• Interconnect the nodes* of the network**

• Connect terminals with the network

• Provide the end-to-end path between communicating processes

Let's now look in more detail into the basic components of a datacomm system, starting with the basic notions behind telephone services.

---

*A point of convergence of communications paths in a network.
**An interconnected or interrelated group of nodes.

19.2 THE PLAIN OLD TELEPHONE SERVICE

The best way to introduce the mechanics of data communications is to review
how the voice telephone works. In the old days, the connection from origin to
destination was established through a human operator. Then, the manual approach gave way to the electronic *selector*. Electromechanical switching substituted the human operator. This temporarily solved the problems of delays
resulting from manual interference, but with the increase in the traffic of telephone calls, bottlenecks were created at the selector level. The conversion of
electrical and electronic selector centers eased, but did not answer this challenge
indefinitely.

Each additional user not only crowds the local selector center, but also impacts
on the whole national telephone network. This is quite important since the demand
for telephone service increases steadily: from the end of World War II onwards,
but most particularly during the last 10 years. Furthermore, from the 1960s, to
the load resulting from voice traffic, are superimposed data communications
needs, served through the existing telephone network.

As stated in earlier chapters, available circuit capacity ranges from 50 to 70
bps for telegraph and telex, to the voice grade lines of 1.2 and 2.4 kbps, the
datacomm speeds of 4.8, 9.6, 19.2 kbps, the newer standards of 56 and 64 kbps,
and up to 400 Mbps for widebands. Furthermore, specialized carriers (the Value
Added Networks, VAN) include as part of their offering, features such as error
control, implemented by the network.

To the individual user, the most important circuit in a network is that connecting his own device, timesharing terminal, or microcomputer, with the network interface; possibly a multiplexer or concentrator. We will be discussing
these building blocks to a fuller extent in the following chapter, but at present
we should clearly state that any communications service consists of three basic
components:

1. *The line of carrier:* twisted wire, coaxial cable, optical fiber, radio bridge,
infrared beam, satellite, and so on.

2. *The switching center:* private branch exchange, local, regional, network
trunk—whether step-by-step, crossbar, or computer-based.

3. *The station at the user site* which is typically a telephone set for voice and
a terminal for text/data communications.

Responding to the demand for network transmission of digital data, new and
established communication carriers have moved to offer facilities which are all-
digital from portal-to-portal. The firmly established trend is toward a digital
network not only for data but also for voice transmission.

The vital nodes in a national telephone network are the Urban Centers: To an
Urban Center converge the short lines. From an Urban Center start the medium
to long distance lines.

An Urban Center houses the line selectors. One approach to design are the so-called *Group Selectors*, which automatically search for free line selectors. This is a step-by-step procedure, selection being done as the user dials the next number.

If the lines are busy, the user has to repeat his number several times, until he gets a free line and establishes his point-to-point connection. This is the procedure of the public lines. A private line is a de facto point-to-point connection, as it links one center (for instance, the computer) to another (for example, the terminal). Using a line discipline, for instance, polling/selecting, we convert this facility to a multidrop line.

A value added network resembles the public line set-up in the sense that the user must dial a number to establish a circuit, but unlike the public lines, the waiting periods in a VAN are very brief and also there exist added advantages such as recovery capability offered by storing the messages intransit in the nodes.

Individual users, who do not justify private lines, have to go through the chore of dialing. This dial-up can be done by a personal computer or a much simpler device. Simple units are now available which memorize a number and repeat it until the connection is established. But a personal computer can offer much more to the user. It can, for instance, register then print:

- the called number

- the calling number (if more than one extension is in the house or in the office)

- the time of the call

- the elapsed time

- and, possibly, the charges

It can also monitor the calls and provide user statistics as instructed.

We said that the line (or carrier) is one of the basic components of a communications network. Let's then take a look into cables and microwaves.

Whenever we speak of the transmission of data over cable, we must bear in mind that the basic circuit is composed of metallic conductors. Attached to the conductors we find some combination of loading coils, equalizers, filters, amplifiers, and repeaters. A repeater is a device used to amplify and/or reshape signals.

A coaxial cable is made up of an insulated conductor tube through which a number of conducting wires are passed. Many telephone, television, and telegraph impulses can be sent simultaneously via such a cable. Coaxial cables form part of the wideband line network. These circuits are usually employed to make distance connections between large switching (telephone) centers, thus providing several independent telephone channels.

The alternate facility is some kind of wireless, radio bridges. Conventional repeater land-based microwave and satellite methods are in use. These facilities may all be described to the network implementor and user in terms of their error rate, bandwidth, signaling rate, propagation delay, turnaround time, and investment.

A microwave is a very high frequency radio wave vitally necessary in handling information because the higher the frequency, the more information can be handled. It is possible to achieve more than transmission of information at these frequencies, as for instance, to perform logical operations such as adding and subtracting.

Transmission equipment, known as radio bridges, uses electromagnetic waves and is particularly suitable for long-distance communications. The number of telephone connections obtainable depends upon the width of the band. Satellite transmission is another alternative.

Both cable and radio waves can be used for private and public lines. By grouping lines together, we can reach higher transmission speeds, if the data load indicates that such a capacity can be utilized full-time. Capacity is the measure of the maximum amount of a material or energy, which can be stored or transmitted.

19.3   SPEED, BANDWIDTH, AND SWITCHING CAPABILITIES

Samuel F.B. Morse's telegraph, invented in 1835, was the only important commercial use made of electricity until 1873 when Alexander Graham Bell began the investigations that led to the telephone. The two transmission networks stayed independent of one another since 1876, and in the late 1950s a third was added to telegraph and voice: that of data communications. (On March 10, 1876, the famous first telephone call took place. Bell called his assistant: "Mr. Watson come here. I want to see you.")

Data communications and voice use in very large measure the same basic components. Telegraph signals could be transmitted over the same network, but for the difference of speed of transmission. Reference has been made to the use of telephone lines.

One important description of data path (including terminal, modem, and transmission line) is the maximum rate at which it can transfer information. It must, however, be clearly stated that raw transmission capability is not the same as the net rate at which information is transferred. There is always an overhead.

On several occasions, we have used as a data transfer measure, the bps. A more meaningful, but less used, measure is the transfer rate of information bits (TRIB), which excludes all overhead. Another descriptor commonly used is Baud: the number of state changes per second. If there are multiple states, then each state can represent more than one bit; therefore, the bps is greater than or equal to the Baud.

For example, a transmission line capable of supporting four states has a bps rate twice its Baud rate, given that each state represents two bits. (Reference to this is usually made as di-bits, and the process is known as phase modulation, but we will not be concerned with it in the present text. (That concept has already been introduced when we said that Baud is a figure indicating the number of times per second that a signal changes.)

There is a correlation between transmission speed and bandwidth. Bandwidth is a continuous sequence of broadcasting frequencies within given limits. To obtain a greater transmission speed, it is necessary to have a wider band. Bands are measured in Hertz (Hz, cycles per second).

Telephone lines normally allowing a speed of 300, 600, 1200, 2400 Bauds are voice grade and have a bandwidth of between 3000/4000 Hz. (The 1200 Baud correspond to the ''normal'' telephone lines. The transmission speeds quoted are indicative; in fact, it is possible to reach 2400 Bauds, provided the line has fairly high quality characteristics.)

From a telephone channel, with a bandwidth of 3000/4000 Hz, 24 telegraphic channels are available with 50 Baud, through the subdivision of the band for telephony. (The Telex service uses low speed, low cost telegraph channels, at 50 Bauds, available to the subscriber 24 hours per day.)

Let's recapitulate: bps is a figure indicating the true bit data transfer rate. Baud is *the* unit of communication. For speeds up to 2400, BPS and Bauds are (incorrectly) used interchangeably. For speeds over 2400 bps, the units of measurement for data transmission and line transmission are no longer interchangeable.

A key element in a transmission network of whichever type, is switching: connecting the lines of two subscribers who want to talk, or two computers who need to communicate. During the first years, this was handled in central offices by a corps of operators using cords, plugs, and jacks. In 1884, Ezra T. Gilliland devised a mechanical system that would allow a subscriber to reach up to 15 lines, without the help of an operator. In 1891, Almon B. Strowger, patented a dial machine connecting up to 99 lines.

Today in America, a telephone can be connected to any of 180 million others. There is a low two-digit number of quadrillion (million billion) possible connections. Though this is a very large number it is still not quite enough for all possible voice, text, data, and image interchanges.

Once again, let's return to the fundamentals. Data communications can take place through:

- private (dedicated, direct, rented)
- public (commuted or dial-up) lines.

Private lines comprise a network, which belongs to the company using it. The solution is costly but provides a faster service in setting up a line and choice of transmission speed. Public lines are hired out by public or state organizations. With public lines, it is necessary to utilize approved transmission equipment, whereby transmission speed is governed by preset choice.

As a matter of fact, there is no hard technical limit, except that quality decrees a level of band usage beyond which the error rates would increase. That is the main reason why telephone companies advise keeping within certain limits, while the use of intelligent modems, for example, is subject to financial constraints.

When we speak of switching, we necessarily mean passing through the exchanges of commutation. Commuted (dial-up) lines trigger off physical connections between two points. Such a connection, set into action via normal apparatus, takes place whenever data is to be transmitted, and terminates at the end of the transmission.

The private lines, too, pass through the switching center, but they are dedicated to a one-to-one or one-to-many (multidrop) connection. The speed is at the discretion of the user of these lines, in function of the modem being used, but the references made in the preceding paragraphs should be borne in mind.

Although the main switching functions for both public and private lines will be assured by the common carrier, local switching processors (frontend) are also relevant to the discussion of communications hardware. For a user to establish a connection with a particular computer, he can dial a *unique* number to make that connection. (The number can also be given to a computer which will dial it by using an automatic calling unit.)

If a user, however, is to dial only one number, or be connected to one leased line, and through that line communicate with many computers (or terminals), then a switching computer must be employed in the network architecture. The switching computer performs the store and forward function, that is, a message is received en route, stored only until the proper outgoing line is available, and then retransmitted.

In one technique, each message contains an identification of the sending device, and an address for the receiving process allowing the switching computer to route messages from any source to any destination. The switching function can also be performed by mapping the input ports of the switching computer to the appropriate output ports; i.e., through a stored table connecting the first to the second.

Besides the inherent flexibility in software switching, the store-and-forward technology allows for optimal line utilization, since interprocess messages can be mixed with other interprocess messages, involving different host computers and terminal devices on the network. The data communications services of the future will generally follow this architecture. And though the inner workings of the system will be transparent to the end user, it is always good that the reader gets acquainted with the general lines of how the system works.

Message switching is another important reference. It refers to the absence of a circuit, logical or physical, relying instead on destination addresses found in the message itself.

It may be preferential to say that with message switching a circuit is established, but only for the duration of a single message. The concepts are equivalent. They allow the switching computers to determine the appropriate path or paths to use, for the message to arrive at the known destination.

Depending on the network, nodes may vary from a very small amount of fixed hardware logic to a larger computer system. Some of the nodes are used only to support the network's connectivity, for example, as message store-and-forward switching computers. Other nodes (sometimes also serving the above function) are the external attachment points for terminals and computer systems. As discussed in the preceding chapter, packet switching presents similar characteristics—with the major exception that it is independent of the intended application.

The basic attributes of a network (that distinguish its architecture) include its topology or overall organization, composition, size, channel types, utilization, and control mechanism (Figure 19.2).

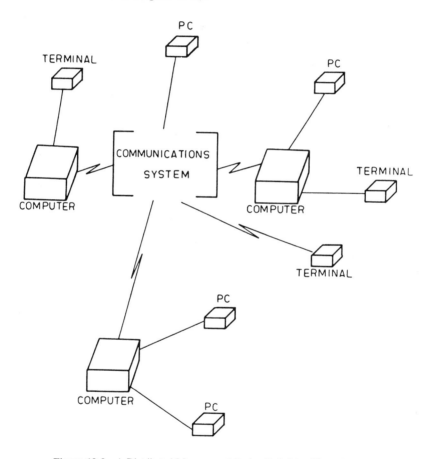

Figure 19.2 A Distributed Message and Packet Switching Network.

19.4 MODULATION, DEMODULATION

A telephone line has electrical characteristics, based on its ability to carry a voice-like signal. When we use voice-grade lines for digital transmissions, we need a medium able to assure that digital information will be carried through.

Modems have been in use to answer this requirement. Their basic function is that of *modulation* and *demodulation*. As Figure 19.3 indicates, there are three basic types of modulation:

1. amplitude

2. frequency

3. phase

Say that a message is to be sent which consists of the bit sequence: 0, 1, 0, 1, 1, 0. A piece of equipment is necessary which would whistle loudly for one second, softly for another second, loudly for two seconds, and so on. This is Amplitude Modulation.

A loud whistle is the code used for 0, a soft whistle used for 1. So a signal can travel through the telephone line. At the receiving end, the computer cannot understand whistles. As a prerequisite for communication, it needs a modem (to be attached to the end of the line), which both understands the whistles and translates them into the sequence: 0, 1, 0, 0, 1, 1, 0.

A Frequency Modulation is best explained as that of difference in pitch. The modem may whistle at the same volume (for instance, always loudly), but at different musical pitches. Here again, data transmission would require another modem at the receiving end, which will understand the difference in pitch, and will translate the received signals into binary digits for the computer. Frequency Modulation has a noise advantage over Amplitude.

Phase Modulation is just as easy to grasp. We have spoken of Hertz (Hz) as the unit of measure: cycles per second. One full cycle has 360 degrees. Each point within the time of a given cycle can be uniquely identified in this context.

In Phase Modulation, the wave form is shifted—each change in line condition may represent a bit. In different terms, each time the line condition is altered, there is a change in bits. For example, the message 0, 0, 1, 0 will have two changes in line condition, from (the second) 0 to 1 and from 1 to 0.

Related to the discussion of modulation and demodulation are the notions of:

• equalization

• scrambling

• conditioning

Equalization consists of telling the modem how to compensate for the characteristics of a particular telephone line. This process is essential because there is a tremendous variety of transmission characteristics that can exist on any given telephone line.

Scrambling becomes necessary for reasons resulting from the computer—or, more precisely, from the string of bits the computer presents as an output. Under normal usage, data processing equipment occasionally produces data patterns:

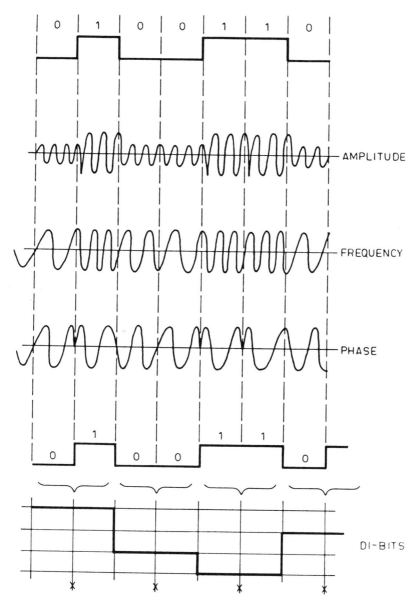

Figure 19.3

like a long list of 1's, that can cause modem problems often centered on matters concerning the clocking circuitry.

A scrambler changes the data, so that it appears to be a random pattern. *A descrambler* is the receiving modem which changes the data back to its original pattern.

Conditioning has to do with tolerances. A tolerance is a statement regarding how far a particular telephone line can deviate from nominal and still be within specifications.

The characteristics of any and every man-made product have tolerances associated with them. Higher levels of conditioning simply *tighten the tolerances*. That is why a highly conditioned line will always meet the specifications of a less conditioned line. Conditioning is therefore reflected in tight quality control, and in that sense, it guarantees the quality of the product.

Having outlined what is meant by modulation and demodulation, we can now speak about the Modem (Figure 19.4). This is the unit, able to change the characteristics (amplitude, or frequency phase) of the electromagnetic wave carrier, transforming the binary nature of the signals which form the characters. Transmission and reception are thus made possible from a distance over voice transmission lines.

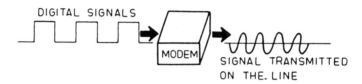

Figure 19.4 Modulation of Digital Signals for an Analog (Voice) Network.

In the modem, the square-edged pulse train* from the computer or terminal, is transformed to fit into the telephone channel frequencies. At destination the signal will again be converted to the original square-edged pulse train. Subsequently, the line controller receives the binary configuration of the characters.

The basic characteristics of a modem are:

- the maximum possible transmission speed

- the types of transmission feasible: Simplex, Half Duplex, and Full Duplex

- the synchronous or asynchronous characteristics

- the time of line inversion: i.e., the time taken by the modem to pass from reception to transmission and vice versa (generally from 20 ms to 200 ms*)

It is vital that the modem is compatible with the corresponding interfaces of the digital equipment: terminal, mini-, microcomputer. Standardization is important and the standard is defined by CCITT.

For short distances or low speeds, it is possible to transmit without using a modem, employing instead simpler and less expensive equipment, namely the line adapter. *Line Adapters* can substitute for a modem, when the distance between two terminals does not exceed 10 miles or when using low speeds: up to 200 bps.

---

*an electrical current or voltage which exists only for a brief period of time and is shaped in accordance with the composition of the message. For instance 1, 1, 0, 0.

*Millisecond—one thousandth of a second (.001 second)

Reference must also be made relative to the way in which a modem works with synchronous and asynchronous mode. Let's call to mind the operational characteristics.

In asynchronous (start/stop) mode, one character is sent at a time, preceded by a "start" signal and terminated by a "stop" signal. Frequency modulation is usually employed. Since the transmission rate is determined by the digital signal, it may vary up to the maximum supported. The existing conventional rates are 110, 134, 150, and 300 bps.

Provision may exist in the character code (e.g., ASCII) for error detection in the form of character parity. Asynchronous modems operating at 300 bps or less are available from many vendors; these are compatible with the Bell 100- series Data Phones. Higher speed synchronous and asynchronous modems are also on the market.

In the synchronous mode, characters are sent in a continuous stream, thus requiring three levels of synchronization: bit, character, and message. Transmission is synchronized by a clock internal to the modem. This fixed rate transmission does not require start and stop bits. A special case of synchronous transmission is the so-called "isochronous," when the clock pulses are given not by the modem (as with synchronous and asynchronous) but by the crystal of the line controller.

To make feasible automatic computer-to-computer communication and, hence, reduce in a significant manner human handling costs, we must make very careful commitments on modems. These concern:

1. automatic call/receive capabilities

2. the choice of the modem itself

3. adaptive multiplexing and other possibilities

In a typical application for the automatic call/receive modem, the computer, say, at the headquarters calls each of the sales offices across the country using an automatic calling unit (ACU) and WATS (Wide Area Telephone Service) lines.

After making the connection, the ACU puts the modem on the line. Price lists and other data are sent from HQ to the periphery; ordering information is automatically transmitted to headquarters from the sales offices. The information is loaded onto disc memory, the computer processes it, and then, using another datacomm link, the plants are given instructions for filling the orders.

Some modems like Western Electric's data set 212 A are called "bilingual," because data are transmitted in two different formats at the two speeds.

• At the low speed, the set operates with a frequency-shift-keyed (FSK) carrier signal.

• The frequency of the carrier signal transmitted over the telephone channel is controlled by the data signal from the customer equipment.

The carrier frequency assignments are such that the modem can operate with other FSK low-speed sets. Data can pass from the terminal equipment to the modem data set in different formats.

1. In the high-speed mode, the data set can furnish timing signals to the terminal, and the data bits must be phase-locked to the timing signals. This is "synchronous" operation.

2. Again, in the high-speed mode, without the timing signals the operation of the set is called "character asynchronous."

With character asynchronous operation the data bits are fed to the modems as characters which are groups of 9 or 10 bits, depending on the type of data terminal.

19.5 THE LINE CONTROLLER

The online connection of a computer to a terminal, and vice versa, requires the presence of a *Line Controller*. Its function is to interface between the lines and the terminals. Its structure ranges from a very simple to a sophisticated one, equipped with a substantial memory and using several line control programs.

Sophisticated line controllers are capable of running the transmission network autonomously and also managing the flow of messages. There are different types of controllers:

• some units only effect a physical connection between the line and the computer, providing an interface,

• others carry out network management functions autonomously, by supplying corrected input messages to the central computer, and by receiving output messages to be relayed to other computers or terminals.

Thus, the line controller is a necessary interface when computer equipments talk to one another over short or long distance telephone lines. The user of a microcomputer, whether calling up the supermarket or his or her club, extending invitations to a social gathering, and so on, will be well advised to understand the functions.

The simple kind of line controller can effect:

1. Data transfers between the central memory and the terminals.

2. The transformation of characters serializing them by bit (we will return to this issue).

3. The opposite operation, giving a parallel (character) structure to serially transmitted bits.

4. The synchronization with the computer for incoming messages and for those to be transmitted.

We must bear in mind that, in order to take advantage of the computer's high speed, it is necessary to have at least a small buffer (for example: one character size) present in the line controller. The exchange of characters between the central unit and this controller takes place through the buffer.

5. The check-up of the received data for correctness (freedom from error).

6. The elimination of the start/stop bits (during the phase of receiving).

7. The transforming code received into the character structure desired by the computer.

8. Housekeeping functions. For instance, during reception and transmission, the line controller signals the "end of text" and the "end of transmission."

In literature, line control procedures are also known as DLC (Data Link Controls). The easiest way to remember about them is they are hardware and software protocols, used for transfer data and control information between separated information devices.

In other terms, to provide a link between separate and often diverse information-handling units, a connection must be established, synchronization of the parties to the exchange obtained, messages passed, and the inevitable errors detected and corrected. A great deal of the standardization in data communication concerns exactly the issue with which we are dealing. (The international coordination in regard to standards is assured by the CCITT, to which participate the public telephone authorities of the various countries. USA is officially represented in the CCITT by the OTP (Office of Telecommunications Policy of the State Department).

Eight technical characteristics help differentiate between one line controller and another.

1. The number of connecting lines. Line control units may be designed to handle just one line or multiple lines. This choice, once made, brings other criteria into perspective:

2. The chosen speed of transmission.

3. The permitted transmission techniques (telegraph, use of modem).

4. The acceptable codes of transmission.

5. The types of connecting line: public (packet switching) or private (point-to-point, multidrop).

6. The number of wires which make up the single lines (2 and/or 4 wires). (An issue to which we will return shortly.)

7. The different types of transmission used: simplex, half duplex, full duplex.

8. The types of transmission (asynchronous, synchronous, packet).

Such choices influence data communications as the range in which these eight design factors we have outlined can vary.

For instance, the number of connecting lines can be anything from "one" to over "one hundred." Regarding codes of transmission, the line controllers can be divided into three main groups:

- either accepting all codes
- only some codes
- only one code

Generally, for a given line control unit, the number of lines is variable, from a minimum to a maximum. The speed of the various lines can either be the same or not. The number of wires making up the lines is 2 or 4. This is important for the different types of transmission, as we will see.

The type of transmission is defined by the number of connecting lines. This is a fundamental aspect which helps distinguish the various line controllers.

As already stated, the different types of transmission are:

1. Simplex

2. Half Duplex

3. Full Duplex

A transmission is called *Simplex* when the data communication only takes place one way. This mode is rarely used in the transmission of data, because there is no return channel for control signals or warnings of errors.

A transmission is *Half duplex* when data communications is two-way, but each "way" takes place at different times. Half duplex requires two wire lines for the transmission of data. Technically, there is also a third wire, often known as "supervisory," working at about 50 bits/sec. This is rarely put to a useful application. A transmission is *Full duplex* when data communication takes place two ways simultaneously. This requires four wire lines.

The selection of duplex, half duplex, or simplex methods will have a direct impact on the choice of the modem. There is a special type of modem for each of the basic types of transmission which we have considered. A four wire modem (two pairs of wires), for example, is required for simultaneous data transmission in both directions. This is, therefore, known as full duplex modem.

What we are essentially saying is that a telephone wire pair is needed for data transmission. The user has the possibility of choosing between three basic types of transmission, but generally speaking, for business applications, the use of a full duplex mode is advisable, even though the software protocol is only half duplex. This is because the reduced turnaround time lowers the amount of dead time between transmissions.

The *turnaround time* is dead time, as far as data transmission is concerned. The signal travels to the central computer resource, is queuing up to use its facilities, is processed by the central computer and an answer is sent in return to the terminal. The time spent travelling through the modem and the lines to the computer or minicomputer, and back, plus the processing time constitute the turnaround.

Let's recapitulate the line control functions: Firstly, the connection between a computer and a terminal is possible only if a specific unit is present: the line controller. The latter is, hence, indispensable in any and every data communication.

Various line controllers are available:

• Units which only effect a physical connection between the data transmission lines and the central computer (or terminal), by means of a suitable interface.

• Units able to carry out the aforementioned functions autonomously, by supplying corrected input messages to the computer (or terminal) and by receiving output messages to be relayed.

The choice is mainly economic, as the technical option does exist. The user must, therefore, decide among the different alternatives, on the basis of the job he has to do and the investment he wishes to make. The simple line controllers must nonetheless carry out the following activities:

1. Effect data transfers between the computer memory and the terminals, or computer to computer, or terminal to terminal.

2. Transform a character, for instance, serializing it bit by bit; or inversely, matching up bits into a parallel bit structure to form a character.

3. Effect synchronization with the computer for incoming messages and for those to be transmitted.

4. Assure that the received data is correct.

5. During the receiving phase, eliminate the start/stop bits and transform the code received into the character structure desired by the computer.

6. In asynchronous transmission, introduce "start" and "stop" bits.

7. During receipt and transmission, relay the end of the message, and so on.

To perform these functions, the line controller requires a number of technical characteristics which are more recently microprocessor supported.

Being an information machine specialized in the transmission and reception of data—and, as such, an indispensable part of data communications, the line controller can be provided as a physical or a logical engine. In either case it can be nicely integrated with the modem and the terminal—as in the last few years we have modems on a chip.

19.6   SENDING AND RECEIVING

To complete our discussion on the fundamentals of data communications, we must consider serial and parallel transmission as well as the sending and receiving phases of operation.

Sending and receiving data over the telephone lines are integral activities of the transmission phase. Transmission can be effected either in a *parallel* or *serial* fashion. More precisely, distinction must be made between:

- Serial by bit
- Parallel by bit, but serial by character
- Parallel both by bit and by character

In parallel transmission, the various bits making up the character are transmitted simultaneously on different lines of the cable, with the same number of wires as bits to be transmitted. Apart from economic considerations, such transmission can normally be carried out short-distance, for example within a building.

In parallel-parallel transmission, more than one character travels at the same time. Here again, the short-distance reference is valid. A, for instance, 16-bit parallel transmission cannot be modulated.

In serial transmissions, the bits of the character are transmitted one after the other on the same media. Serial transmissions are carried out on the usual telephone lines. This represents the regular type transmission. When we count in bits per second (bps) we must always keep in mind this "serial" feature.

Say that a line supports 2.4 kbps (discounting for a moment the difference between theoretical and practical transmission capability), and that we transmit 8-bit characters. In this case of a synchronous discipline, this corresponds to the transmission of 300 characters per second for data and for control characters. In an asynchronous mode, the throughput is lower, since we must also account for the start and stop bits.

Simplex, half duplex and full duplex refer to serial/serial type of transmission. We have already introduced these issues.

Whether we operate in a serial or parallel mode, the functions to be carried out by the data transmission network can be divided by phases. During the *receiving* phase, such functions include (Figure 19.5):

1. Initiation of a communication
2. Control of the received data online
3. Management of timely communication with the terminals (periphery)

RECEIVING PHASE

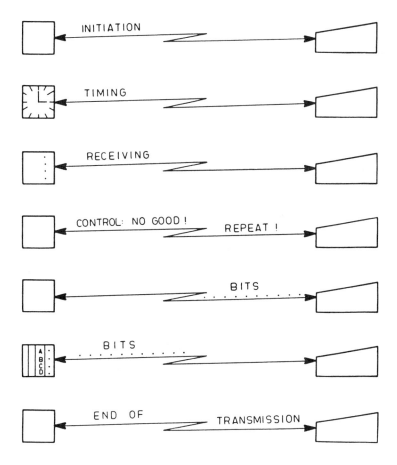

Figure 19.5 Receiving Phase of a Terminal Device.

4. Depending on the type of equipment, reception of bits, and their assembly into characters

5. Assembly of the characters into messages

6. Checking for errors during the transmission

7. Issuing of possible requests for repetition of the message

8. Suppression of bits used only for data transmission purposes. For instance, "start" and "stop" bits, used to indicate the beginning and end of characters

9. Recognition of "conversational characters," such as: "end of message," "end of transmission," and the like

10. Elimination of bits or characters which don't belong to the data, e.g., bits or characters injected for the convenience of the conversation

11. Storing the messages received

12. Forwarding these messages upon request or as instructed

We spoke of "conversational characters." We also said that ANSI defined a base character set, which uses seven bits/characters with an 8th bit parity. A total of 128 characters are thus available, some of them being of daily usage in line control—and we have given examples on them (in the chapter on Protocols).

Let's now recapitulate. There are many functions executed during the transmission phase, starting with the storing of the messages to be transmitted, queuing them up, converting those messages from computer code into various transmission codes, and introducing the needed characters for data communications. This is followed by transforming the messages into characters, breaking up the characters into bits, initiating the message transmission proper, and forwarding the bits online in a timely manner.

Finally, control action calls for supervising the transmission process, with the possibility of repeating messages, in case of error. Giving a warning of the end of text, end of transmission, and the like, are part of the control activity. These operations are commanded by software and executed automatically by the machine—hence, in a manner transparent to the end user.

Chapter 20

REALTIME, TIMESHARING, MULTIPLEXING, AND LINKING THE PC TO THE MAINFRAME

20.1 INTRODUCTION

We have spoken of the fundamental notions involved in data communications and of the role protocols play. With the latter has been associated the most modern discipline: Packet Switching. It is now time to return to the fundamentals and look into the evolution of data handling disciplines—and the steady growth of communications needs.

Online connections are characterized by teleprocessing routines. The term *teleprocessing* is not particular to personal computers. In its original sense, it identified a data processing system where a central computer is linked up to terminal points on the periphery and possibly also to satellite computers via a data transmission network.

The arrangement from the center to the periphery, resembling a radiating sun, is known as *Star*. An alternative is to connect different computers in a *Loop* fashion. Still another, more modern and more efficient, is to use *local area networks* (LAN). A fourth possibility is to employ a PBX (*private branch exchange*) (Figure 20.1).

Whichever structural solution is chosen, data communication can be accomplished either in remote batch or in realtime. With a star, data communications will take place from the remote locations (in the periphery) to the center and vice versa (hierarchical). With a ring, data will travel between these remote locations themselves, in a peer system.

As the name implies, local area networks will connect intelligent workstations in a local area—typically up to 1 km (0.6 miles) for baseband, up to 10 km (6.2 miles) for broadband. Such LAN will be interconnected long haul (wide area) among themselves and with other computer resources through gateways.

The PBX act as gateways. Computer-based units help integrate voice and data, and can be used within the same building to interconnect terminals as well as voice stations.

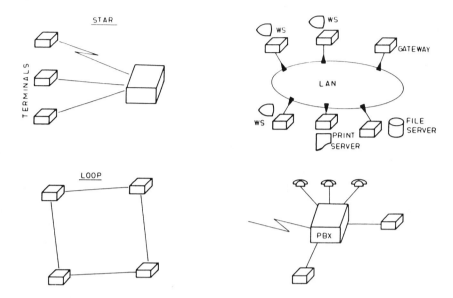

Figure 20.1   Interconnecting Computer Networks.

Within this general description, we distinguish different ways in which the line connection will be made (See also the discussion in the preceding chapter). In a classical realtime system, each terminal point is permanently linked to the computing center, usually through private telephone lines (This type of connection prevailed from the mid-1960s to the early 1970s). In a remote batch system, the connection is assured in a periodical fashion through public telephone lines.

Timesharing will require a permanent connection to a central resource or computer utility. Its first implementation took a star form, as all terminals (and the users employing such terminals) have been connected to a central resource.

A ring connection often characterizes a network of independent processors, possibly minicomputers or microcomputers, communicating with one another via transmission lines, none having an overall supervisory faculty over the others. This is the form of a *distributed* system, without centralized control functions.

Typically, a distributed information system consists of two or more nodes, which may share capabilities and exchange messages via communication facilities. The control functions may be centralized or decentralized among the nodes of the system. A LAN is a distributed system, and so are all peer-type structures.

Teleprocessing is the key to distributed systems. What is really the sense of distribution? To answer this question let's go back to the fundamentals. In an online operation, the main functional components are the computer, the line, and the terminal.

Communication lines are an integral, indispensable part of an online system. But they are not the only one. A computers and communications network has many components. Some of these are physical. Others are logical.

Interconnecting the physical components through logical media calls for tele-processing. In a modern environment, line disciplines are not so much necessary to share computer resources. Each user has his own, dedicated *personal computing* facility. Teleprocessing is the key to sharing databases and operating electronic message systems.

20.2 BATCH AND REMOTE BATCH

By batch* processing is meant a way of handling data, characterized by the fact that processing is done by blocks of data, accumulated over a given period of time. The preparation of a payroll, billing, and accounts payable are examples of batch processing.

A number of data collection points exist on the periphery. The data accumulated there is forwarded to the center by auto, train, air, and other media, on the basis of a daily, weekly, monthly, or other time interval. This information is sent to the computing center, together with other updating references, such as—additions and deletions from payroll, new customers, and expediting orders.

At the center, the data on the original documents is transcribed onto a support, readable by the computer. In this way, the computer acquires the needed information for processing: updating of files, calculation of bills, preparation of payrolls, and so on.

The first operation done by a computer in a batch processing mode is data clearance: that is, the assurance that the received information contains no errors. Whenever errors are showing up, corrective procedures must be put into effect to weed them out. For instance, during the phase of computation of the payroll or customer billing, management can obtain some byproducts, such as statistical data.

To recapitulate, batch data is handled in blocks, and each block is treated sequentially through a defined procedure. Furthermore, each procedure is *usually independent of the other processing procedures*, though it may, for instance, update their files. Batch processing is characterized by:

1. The accumulation of data over a certain period of time. This data will constitute the *input* of a data processing operation.

2. The division of data handling into successive, well-defined phases of processing: key to disc, checking for errors, computing, and storing.

3. The creation of an *output*, either in a printed form or stored on a support medium.

While the accumulation of data on the periphery may continue to be carried out through manual methods, the transmission of data to the center might be

---

*Precisely, the quantity of anything made in one operation or lot.

performed by means of a telephone line during say, evening hours to take advantage of lower rates. In different terms, data will still be collected in blocks and kept locally over a period of time (hour, day, week), but they will also be locally transcribed onto a support (tape, floppy disc, and so on) and then transmitted to the center for data processing.

This procedure is better known as *Remote Batch*. It is the first step toward online operations.

Quite likely, for data processing operations which take place in business, industry, and banking, persons faced with communications problems may prepare a quantity of data to be sent to another party through regular mail, air-mail, or telephone lines. Typically, the things to do or to say are stored today in human memory or, as the case may be, a note is taken by paper and pencil for future reference. But notes can often get misplaced.

It is much more of an elegant solution to accumulate the lot of data on an inexpensive personal computer, then forward this data to the party concerned at the right time. Commonplace in the future, this will be the case of a person preparing a shopping list, which will be forwarded to the supermarket computer to computer by remote batch.

Instead of standing in front of a counter for a long time, waiting to get his airline ticket, the business person would use a personal computer to schedule a trip and identify the options available. He would then transmit the schedule, computer to computer, awaiting confirmation, and for the ticket to be printed on his/her private terminal.

The race between communication and transportation is just starting, but communication should not only be viewed as being exclusively of the medium- to long-distance variety. Under present conditions, communication within a building can take a day. Simplifying the task of writing memos and reports can have a tremendous impact on the lives of the individuals and of that of industry. This brings under perspective electronic mail. It is a protected message switch capable of distributing a document to multiple terminals (video and/or hardcopy).

This is one of the implementations to which is put the concept of "a terminal on every desk," and it is just as likely that electronic mail and voice mail will be part of the new telephone service, equipped with a videoscreen and a simple keypad. Another implementation is videotex.

Rather than spending hours signing checks and writing out envelopes to be expedited, the consumer would order his bank to do so via his personal computer, intelligent telephone set, or TV screen, connected to the classical voice network—or a value-added carrier. In doing so, the consumer will work realtime.

20.3   REALTIME AND TIMESHARING

Realtime presupposes an online connection: for instance, a telephone line which links a device at the disposal of the user to a central computer resource. To better explain this concept, we must first make a flashback.

Whether we speak of batch or remote batch, we make reference to the offline preparation of blocks of data, which must subsequently be passed on to the computer. Offline is the opposite to an online operation. The local data collection point—a sales office, a factory, a banking branch or a home—is equipped with, say, a microcomputer but has at this particular moment no telephone line connection to communicate with the center. It may, however, dispose of that faculty at a later hour, a case usually referred to as an "offline/online" type of operation.

In a typical offline operation, the available equipment is used for data collection. This is advantageous, as the local equipment can help weed out error. But between the first information collected and the last, there is a time interval, or delay. This is a multiple of the time it takes for a block of data to be sent from the periphery to the center.

In many cases, the time interval is perfectly acceptable. In a payroll situation, weekly data is sent to the center to be treated at the end of the month. But in other cases, delays of say a day or two are acceptable, but longer delays have a financial impact on the users' side. It is the case of accounts receivable. When bills are sent out late, collection is also delayed.

For some procedures it is, therefore, necessary to reduce this waiting period, and one has to examine the cost of the faster dispatch, along with the benefit which it brings. In still other cases the delays inherent in batch processing involve a risk. A bank honoring a check needs to know the balance available in the client's account to avoid overdrafts.

A direct access to computer memory thus becomes a sought-after facility. A procedure of this nature characterizes realtime operations. By definition, realtime is the direct access to computer memory for fast updating of a file, or to retrieve a small amount of information. Realtime is:

- no batching-up of data
- treatment of each piece of data as it comes, when it comes
- the possibility of correcting errors online
- handling data case by case
- simultaneous processing in many cases
- access to all files available online

Since realtime involves online access to a database, such access must be protected both at the level where the information elements are stored and also in the transmission phase. The former is done through authentication and authorization—the latter by means of encoding and decoding.

Channel access can, thus, be made safer through the transmission and reception of encoded data. Encoding and decoding is effected either through hardware (such as the 64-bit chip of NBS) or in a software sense. We have spoken of these possibilities.

Let's recapitulate. Realtime means a quick access to a file to check something in the database, or to update a given file. A good example in this direction is the cashing of a check. A client presents himself at the teller window to cash a check of, say, $300. The teller does not know whether the client has sufficient funds to cover this check. He can interrogate the files, and this can be efficiently done online. Two operations are involved:

(a) Is the available balance greater than $300? A "yes/no" answer will be given by the computer. If "yes," then:

(b) Subtract $300 from available balance.

(c) State new balance.

Inventory management and airline reservations are other examples, characterized by the fact that a client presents himself, asks for a service, and must receive as fast an answer as possible. The time the system takes to respond is known as the *response time*. The time the central machine takes to calculate is the *turnaround time*.

There are evident differences between remote batch and realtime. A critical question with computers and communications has always been: "Is the data transmitted from the center to the periphery still actual?" Actuality is generally assured in an online situation, while with batch, in most cases, by the time the printed output returns to the periphery, the information is long obsolete.

Instantaneity is, therefore, a basic difference between remote batch and realtime. Yet, when we speak of fast or instantaneous response, it is proper to identify what we mean. In a banking or airline reservation system, response may need to be at the level of about two to five seconds. To the contrary, an inventory control system may have a response of a substantially longer time period.

There is no universal definition of an instantaneous response time, that time varying with the application and the procedure. Another basic difference between remote batch and realtime is error detection and correction.

A fundamental characteristic of realtime response is the assurance that the errors have been weeded out prior to tackling the files. The procedure is known as error detection and correction. In a realtime situation it is done online, and hence the user is assured that only a minimal amount of errors are left in—in such a way that it will not disturb the processing operations.

Error-free operations and simultaneous response are the two main reasons justifying the cost of realtime. To the contrary, with batch processing much erroneous data can persist, and the first duty in data treatment by central computer is to weed these errors out. In conclusion, the following characterize the realtime procedures:

- computation immediately follows data input

- the incoming information can reach all relevant files, updating accordingly

- this requires direct access memory devices, enabling a speedy search, of for instance, disc files

Both input and output are (usually) transmitted over telephone lines, satellites or radio bridges, and this brings into perspective the issue of data communications. When many users refer to the same computer resource, this operation is characterized by the sharing of the resource in question, which naturally leads to the subject of time-sharing.

Literally the term *timesharing* means division and sharing of a common time, but to be more precise it's preferable to speak of *division and sharing of common resources*. This concept, applied to computation of data, defines a system which can be of service to users, the same way a utility company serves its subscribers with power.

In timesharing, a central system (whether mainframe or minicomputer) is at the disposal of a number of subscribers who use its resources via terminals, installed at their workplace. Such a concept can thus be likened to a common service, which under certain conditions, can resolve the user's problems more economically than if each had a central equipment of his own.

Technically, timesharing involves:

- data input at the user's site
- data communications
- access to central files
- computation and response to the computation
- output printed or on a videoscreen—at the user's site
- the employment of common computer resources

What happens with timesharing is this: When several people—Mr. BLACK, Mrs. WHITE, Miss BLUE—put forward requests at the same time, the computer goes ahead with the first user for a fraction, whilst making the second and third user hold on. Then the computer lets the first and the third user wait, and deals with the second for a moment, and so on.

These waiting periods are so brief that the users themselves are practically unaware of them and, in fact, carry on as if the computer were at their disposal. This is true up to a point. As the common computer resources get crowded with requests response time increases. Hence, the wisdom of using *personal computing* principles through PC.

Queueing up for computer servicing is quite like queueing at a supermarket's cash-desk, or at a medical service center. Typically, there is a bottleneck. The available resources get outpaced by the demand for them, and that is why the people making the demand must wait their turn in the queue.

Messages, too, must wait in the line, one after the other, until the computer facility (service center) is able to deal with them. A logical way to look at cost-effectiveness would involve the following steps:

1. Examine the cost of the computer and the system load to be applied to it.

2. As the number of users increases, cost to users decreases.

3. Past a certain point, the central computer resource becomes very complex, the cost per user increases again.

4. At the same time, as the cost of the basic electronics sharply decreases, it becomes more economical to have a microcomputer dedicated to each user.

A different way of making this statement is that timesharing has its limitations. To serve many subscribers, often with diverse interests, the computers must have multiprogramming capabilities and impressive storage media. That is, systems geared to complex operations.

The user must pay for the terminal, the communications line, and his share of the central resource both computational and memory (for storage). The bill can be impressive. Personal computers have opened new horizons in assisting each person with computer power. It is an opportunity which should not be missed.

20.4 MULTIPLEXING AND CONCENTRATING

One basic issue with network communication is the cost advantage of high capacity communication links. But a mechanism must be provided to share these facilities among multiple users. Multiplexing is one solution—this term being generally applied to non-programmable devices. Let's see the possibilities:

• Frequency division multiplexing (FDM) lets a number of low-speed devices share a link, by dividing its frequency spectrum into a number of subchannels.

• Time division multiplexing (TDM) assigns each low-speed device a time slice into which it may place information. TDM multiplexers operate on bit-interleaving and character-interleaving modes.

TDM can allow a notable reduction in the network costs (especially in medium-sized system configurations) by the fact that the time division multiplexer gives access to a certain number of lines from the one side, and from the other, restores the bits received to one single line in a specific and strict sequence.

Such a synchronization requires the line connecting the peripheral system to the computer to support a speed, which is at least equal to the top speed borne by the lines entering the peripheral TDM, and those leaving the devices connected to the line controller. For instance, if the lines under examination are 1.2 kbps, the line connecting the peripheral network to the central system should be at least 9.6 kbps.

The use of such equipment does not make for significant increases in reply time, and therefore does not deteriorate the system's performance. Nevertheless, the lines connected to TDM equipment are continually linked with the central system and must be dedicated (private) lines.

An example of frequency division multiplexing is found in undersea telephone cables. Before such a large number of signals can be dispatched from a land terminal to an ocean cable they must be processed in several major steps. In the first multiplexing step, 16 signals are combined into a "Group" by modulating the individual three-kilo-hertz signals into adjacent frequency bands.

In the next step, most of these groups are combined, five at a time, to form "Supergroups" of 80 conversations each. Ten supergroups form a "Master-group" with a bandwidth of 2,520 kilohertz (thousand of cycles per second).

Six mastergroups (Bell's L4 System) are assembled into a "Jumbogroup" for transmission through a single coaxial cable at a frequency of between .5 and 17.5 megahertz. The still more advanced L5 system is able to transmit three jumbo groups in a single coaxial cable at a frequency of between three and 60 megahertz.

It is appropriate to distinguish between multiplexing and concentration, two terms often used synonymously. Multiplexing generally refers to static channel derivation schemes, in which given frequency bands or time slots on a shared channel are assigned on a fixed, predetermined (*a priori*) basis. Thus, a multiplexer has the same instantaneous total input and output rate capacities.

Concentration, by contrast, refers to sharing schemes in which a number of input channels dynamically share a smaller number of output channels on a demand basis. Hence, concentration involves a traffic smoothing effect, not characteristic of multiplexing.

A concentrator is, generally speaking, a programmable device. A multiplexer is not. Since the aggregate input and output bit rate need not be matched in a concentrator, statistics and queueing play an important role. What the two solutions have in common is that both multiplexing and concentration are means for obtaining a cost-effective teleprocessing network, which is postulated on efficient utilization of the communications links and processing equipment. Furthermore, the declining cost of small computers has made it attractive to replace hard-wired multiplexers with programmable processors.

The inherent flexibility of computers made other functions possible: By using small processors as the concentrators, the dialogue between these and the frontend processor may become more efficient. In turn, this can improve the efficiency of the network, by permitting more sophisticated line control procedures which may also facilitate "recovery from failure" situations.

Typically, some of the functions previously performed by the main computer, or the frontend processor, may be delegated to the concentrator, giving a more immediate response to the terminal. The polling of local lines, or resolution of contending terminals, the checking and formatting of messages, the correction of errors, guidance to operators and similar tasks are representative of those conveniently performed by a concentrator with a reasonable degree of processing power.

As technology made such developments possible, it is appropriate to use the term *concentrator* to specifically identify a programmable unit with at least some memory capacity, basically able to effect store-and-forward operations. The buffer in the concentrator will thus be employed to accumulate characters coming from a terminal, until a terminal condition, such as buffer overflow or an end-of-message segment signal, is received.

The accumulated message will then be transmitted over the high-speed network circuit, taking advantage of the available capacity. Error control procedures and message reassembly may be implemented in the concentrator, as well as terminal speed recognition and code conversion. The concentrator may reform the data, in order to interface different speeds and protocols.

Adaptive multiplexing or bandwidth contention takes advantage of the statistics of circuit usage. This technique provides a multiplexed channel on a demand basis for each individual circuit. The multiplex channel exists on an end-to-end basis only when the terminal-to-computer circuit is active. Otherwise, that portion of the composite data stream it occupies when active is available for other circuits.

The end-to-end multiplex channel is provided on the basis of control signal indications and specific channel request data characters. The outstanding characteristic of adaptive multiplexing is that the multiplex channel is dedicated to the terminal-to-computer circuit during the period of its activity.

Activity on other circuits through the multiplexed path will in no way delay or disturb it. Adaptive multiplexing is, therefore, a nonblocking mode of transmission. Because the channel is dedicated and since there is no data manipulation, the adaptive multiplexer is completely code-transparent and can transmit all the standard code sets and also random data patterns such as those of the National Bureau of Standards encryption system.

Propagation delays in adaptive multiplexing are relatively short. The delay is fixed so that the data transmission does not take on a bursty characteristic. The proportion of the high-speed data stream occupied by the channel when active is also fixed and is either equal to (for synchronous channels) or less than (for asynchronous channels) the data rate of the circuit.

During periods of peak traffic, active circuits may occupy all of the available high-speed bandwidth. Other circuits which request a channel are either put in a camp-on queue, busied-put, or not answered, depending on network application.

20.5   FRONTEND PROCESSORS

In network communication, many of the tasks performed are relatively simple, but highly repetitive and can make considerable demands on the time of a computer. It often proves more economic to perform these functions by a small separate computer called a frontend processor (FEP). The main advantages of this approach are:

1. The cost of hardware to attach lines is often less with a small computer, possibly because different constructional methods are used.

2. The processing load removed from the main computer will considerably increase the power available for computational purposes.

3. It becomes possible to separate the complete system cleanly into two parts: the main processor and the communications network.

This gives increased flexibility and may allow one part to be enhanced or replaced, without affecting the other. To recapitulate: Whether using concentrators or frontend computers, each terminal is directly linked to this unit, situated in the best suited place for the terminals themselves, from a functional, technical, and economical point of view.

Frontend computers are advisable when there are many terminals spread through the same building; in different locations in the same city; or in different cities, near to one another, but all far from the center. The frontend will receive messages from different terminals, and retransmit them to the central computer (or another peripheral minicomputer, in a hierarchical structure) by means of a one-way communication.

Messages arriving from the computer are treated in the same way. As with concentrators, the speed on the line connecting the frontend with the computer is normally higher than that found on the lines linking the different terminals to the frontend.

Technically, both concentrators and frontends have some aspects in common with the line controllers—which we discussed in a preceding chapter—as they must be capable of:

- conversing with the various terminals
- providing collection and re-distribution of messages
- transmitting and receiving service and control characters
- speeding the traffic of messages between the line running to the computer and the lines going to the terminal
- seeing through a possible code conversion (for instance, receiving a code from one terminal and transmitting it to the computer in another code)
- guaranteeing error control procedure
- memorizing messages in transit (store and forward)
- dealing with automatic dialing operations and procedures

The principal characteristics of evaluating the capabilities of a concentrator or frontend are:

1. the number of entrance lines

2. the type of transmission (synchronous, asynchronous)

3. the speed of the incoming and outgoing lines (terminals-concentrator-computer)

4. the ability to memorize data received (presence of buffers, memory, storage media)

The major advantages of frontend equipment are in off-loading the central computers: processing memory, and storage facilities, and in providing flexibility in the number and variety of devices that can be connected.

In performance of these tasks, the frontend is responsible for: line control of all communication circuits between the central computer and remote terminals; the control of the interfaces, typically through a memory channel, to the central computer; message formatting to translate messages to a format compatible with the destination computer (or terminal); error detecting and correcting to guarantee the integrity of transmission; device multiplexing to assure the efficient use of channels and circuits; and device control allowing for flexibility in the connection of different peripherals to the central facility.

Summing up, a concentrator or frontend functions in a continuous manner, as a center for receiving and transmitting messages between distant terminals and the central computer. By interleaving frontend (or concentrator) in a data communications network, it is possible to optimize the configuration: number of terminals, geographical aspects, bottlenecks, waiting time, addressing procedures, and so on.

However, the simple introduction of a frontend will not bring automatic benefits. The degree of cost/effectiveness to be attained with computer equipment is always a function of the long, hard look taken in the course of a systems study, and of the steps taken to implement a systems solution.

20.6 MEMORY TO MEMORY

The personal computer can call a long, for instance, 9-digit number (function) on the basis of a brief 2-digit code (argument) given by the user. This argument-function combination may identify another party with whom the user wishes to talk by voice, but it may also be a computer-to-computer communication.

A user, for example, may order through a PC what is needed at the supermarket, while sitting at a terminal at home. The items are keyed into the terminal as they come to mind and displayed on the monitor (videoscreen).

This terminal, the family microcomputer, keeps the data in question in storage. Then, by midday, when the list is ready, the user commands the machine to transmit memory to memory to the supermarket's computer, for immediate delivery.

This application can go further. If the bill exceeds the budget the housewife has established, the computer at the supermarket compiles the costs and transmits them to the house (personal) computer for authorization. This way, by using the communications links and their own software, the two computers engage in an interactive exchange of messages.

Another person wishes to send a message to a friend or to a community group, whom he/she knows to have left home at a particular time. The message is keyed in, stored in the computer, and the machine is instructed to transmit it to the memory of the other personal computers. The person will do the work only once, then the computer will transmit the message, once per person, to the parties concerned. A housewife can send a message to her husband, paging him in a meeting, and delivering the message on a video (softcopy) or in printed form (hardcopy). This way her husband does not have to leave his meeting.

All this is likely to end up costing less than the bill to be paid for normal telephone calls, because the data communications solution is so much more efficient. Efficiency has often been one of the issues where machines tend to excel over people.

In the business arena today, there is an application which gives a fascinating glimpse of what some of the communications technology can mean in managing a flood of office documents.

There, each letter, report, memorandum, and contract is recorded on updatable microfiche as soon as it arrives in the mail room.

• A comparing record is made in a digital file, noting the origination, destination, date of the document, etc.

• Thereafter, the images which have been selected for this control can be retrieved by one or more of a dozen variable terms.

• The digital system identifies the microfiche (and the panel) on which each image is stored.

The information can then be viewed on a reader, or reconstituted as a full-size hardcopy.

So much for historical records. Active files will be better handled through a computer-based automated environment, where a manager and a secretary create, transmit, receive, and read information via video terminals. They file their documents on discs. Such microcomputer-based workstations can easily communicate with each other, by means of dial up telephone lines, thus creating a complete integrated data handling network.

To recapitulate, the critical components in data communications are: the terminals operator, the terminal, the local computer power, the units needed for line control, the carrier (transmission lines), the files (at the local computer(s) and, if any the central computer), the "other" computer resources—local and central, logical, and physical.

The personal microcomputer may be used in this connection as a simple, easy to use, yet powerful workstation, calling up a number, passing on a message, playing a game. Or it may be used as a more sophisticated unit. In either case, it will be performing a similar function in the home, to those it often performs in an industrial environment.

After all, the telephone as we know it today is not much different in terms of end usage, whether it is installed in the home or in the factory (and likewise the radio).

Still, it is correct to bring under perspective that communicating computers (micro, mini, or mainframes) need more supporting facilities with which to work. If they are going to do computational jobs and also communications, there must be an interrupt service to make this duality possible in a manner transparent to the end user.

Slightly more sophisticated functions than those able to be served through a one-track type of handling require the existence of an interrupt system. This makes feasible concurrency: Following an automatic signal, an urgent procedure goes into operation, interrupting what was going on at that time—given the speed of the computer, this is transparent to the end user.

For example, computer 3, a node in a network, may be programmed to give first priority to messages from the other nodes/switches; second, to messages from the computers which it controls; and the third to the terminals directly attached to those machines.

A higher priority can interrupt a lower priority job, and assume control of its resources. But the computer must keep in memory what was going on at that moment in order to be able to return to it. In general, we can distinguish at the local site:

1. Input/output interrupts

2. Processor interrupt, devoting the resources to a specific condition

3. Interrupts concerning specific program issues

The word interrupt brings up the need for priorities. Even a very small computer, or an intelligent terminal, can be used for this or that function, on the basis of pre-established requirements. As in the example, which we gave, certain functions, when they happen, are allowed to master the available resources at the expense of other functions which were going on at that time.

The so-called Input/Output (or External) interrupts enable outside interventions from the operator, the end user, a given process, a remote terminal or another computer. For example, the end user at the console may introduce new work priorities, carry out interrogations, put a particular procedure into effect, and so on. Quite similarly, another computer may send a signal obtaining priority over the use of the resources.

The user may be in a routine type conversation, using electronic mail facility, but the school master (if he has a priority status) may flash-in to inform the user that a child is sick and must be fetched immediately. The user may use the computer for betting with friends on a baseball game, but a boss (if he has a priority status) may take control of the resources of a personal microcomputer to ask questions about a conference held in the morning or to prepare one for tomorrow.

Another class is the Internal interrupts. Through them, peripheral units are able to function autonomously, in conjunction with the minicomputer, signaling when an operation is finished, or giving a warning when an anomaly occurs. The main processor unit may come into play at the appropriate time, checking out:

- a completed operation
- the start of another operation
- the unavailability of a given peripheral unit, and so on

intervening in such a manner, as to unaffect the continuous operation of the peripheral units, with the exception of the one whose current activity must be interrupted. Or, an internal clock may interrupt the main processor unit to calculate a time interval, to memorize accounting data, or to single out data processing cycles.

Interrupts may result from hardware or software causes. The computer may have the faculty of automatic interrupt in the execution of various programs which give way to errors, or make it impossible to proceed, because a given instruction has not been followed correctly.

The occurrence of irregular conditions—i.e., other than those specified by the program—is another reason for interrupts. Program-based interrupts bring the main processing unit into play. Control takes corrective action, or signals to a higher-up authority the impossibility of continuing with a certain process.

Without this facility, the computer would have carried on regardless of an error contained in the program or the data. Indeed, the ability of the computer to identify its own errors and take the proper corrective steps is one of the vital faculties which simpler devices do not have—though they may be doing some of the lower order functions the computer also does.

20.7 CONNECTING PC AND MAINFRAMES

PC are spreading throughout corporations, but much of the data needed are still in the mainframe. Physical and logical connections must be provided to the user.

The *physical* connection is typically done through an RS 232C cable. More complex is the *logical* connection which must consider the type of work being performed:

1. Using the PC as a dumb terminal
2. Hard disc to floppy transfer
3. Hard disc to central memory (of the PC)
4. Communicating processes at the applications level

Stated in different terms, the PC cannot exist in a vacuum. Datacomm packages are necessary to support terminal-type usage. Simple protocols are: TTY ASCOM, 2780/3780, 3270 BSC. An architecture bound protocol is: 3270 SNA.

Third party sophisticated protocols are: Visilink/Visianswer, Data Connection, PC Express, Peachlink, and lots of others. A standard format is also necessary, such as: DIF.

The simplest link is to have the personal computer *emulate a non-intelligent terminal*. This is the way most computer users interact with commercial databases. But such a connection only allows the PC to look at the data, not to store them and manipulate them.

One step up is *the ability to download data*. This permits the end user to call information from the mainframe and store it. The data are in raw form and software is necessary to get from storage into, for instance, the spreadsheet.

An improvement is to *add software to allow the data to be formatted*. So, they can go directly from the mainframe into the proper slots in the desired program. Another approach is to *have the WS and the mainframe run the same programs*, so that translation of data is not necessary.

Mainframers, PC vendors and software companies are interested in providing this interconnect facility. IBM introduced engines designed for this link: XT/ 370, and 3270 PC. The 3270 PC doubles as a computer and as a 3270 type terminal.

Interconnect facilities also pose other challenges. Security must be preserved, and there must be strict controls to make sure that errors are not introduced into central files. *Connecting PC and mainframes* is the way to look at information systems design. Different reasons point to this arrangement.

First, the satisfaction of the growing ranks of corporate WS users. These users call their need to tie into the mainframe: *critical*. Second, the reduction of processing loads on the mainframe. Third, with intelligent WS, the processing job is brought where it belongs: *at the workbench*.

Fourth, the *tailoring of the application* itself to the individual needs of the user. Fifth, the benefit in considerably reducing *response time*. This is achieved because *computer power is now dedicated to one job*, and therefore has the ability to do it well. We can only achieve an effective solution if we integrate PC and communications. This permits us to use the data which is already there in private and public databases. And it is the key to productivity increase. ''Island solutions'' have little effect on management efficiency. The happiest user of standalone PC, in three to four months will not be happy anymore. He will require online services.

The subject of connecting the personal computer-based, intelligent, interactive multifunctional workstation is so important that it requires a full chapter in itself. Chapter 21 is dedicated to this issue.

Chapter 21

GOOD NEWS AND BAD NEWS ON THE PC-TO-MAINFRAME LINK

21.1 INTRODUCTION

The successful implementation of intelligent, multifunctional, interactive workstations is tantamount to their online connection to mainframe resources, particularly to large databases. The independent, standalone, noncommunicating WS is of little or no value in the longer run.

The professional workstation may work standalone part of the time; but this work will be so much more rewarding if the WS supports, in a network-wide sense, file import/export and communicating processes. This reference alone documents the wisdom of answering the question: "Why the link?" in a positive sense.

Like all technical issues, PC-to-mainframe connectivity poses challenges. They can be phrased in four points.

1. Consistency in terms of solutions

2. Establishment of an efficient, error-free data link

3. Ability to integrate at a higher level of the ISO/OSI model

4. Obtaining solutions at reasonable cost

The approach which we will take may reflect either level of connection, depending on requirements. As discussed in the preceding chapter, these levels range from the simplest, emulating a non-intelligent terminal to a more professional, including downloading and file-to-file transfers. The next level is formatting prior to downloading and to upline dumping. Finally, a more sophisticated approach is process-to-process communications.

Whichever alternative is chosen, it must make a system solution. Such a solution will typically go beyond the usual prerequisites to an online communication: a modem, a line, a chosen communications protocol, and the need for formatting.

A valid system solution will include decisions throughout the seven layers of the open system interconnection, all the way to document content and document interchange characteristics. We must get ready for the day when computer conferencing and full blown office automation will become the order of the day.

While our perspective in PC-to-mainframe connection should focus on the highest level of systems architecture, the first choice we must make concerns the lowest layer: the modem. Today PC communications are usually supported by low speed modems. This will not last long.

The 1984 statistics indicate that only 8.8% of PCs being sold included a modem in the configuration. The forecast for 1987–88 is that this share will grow to 53%. The same forecasts indicate that by 1987–88, the 300 baud modem will be a relic. Demand will focus on modems working at 1.2 to 9.6 KBPS.

A user organization will be well advised to start with these modems now. Furthermore, professional PCs should be used with automatic modems which are characterized by automatic dialing and answering, fully unattended operations, and programs control.

PC-to-mainframe communications should support logon procedures in sending/receiving computer files. It is also advantageous if the modem features forward error correction to improve the bit error rate (BER) of the lines.

Forward error correction (recommended standard V.32 by CCITT) reconstructs the bit/byte sequence without requiring retransmission. It costs about 20% more in terms of line usage (allocated to polynominals for parity checks) but helps improve line quality by two or three orders of magnitude.

There are other goals we may wish to reach at the modem level. Relative security is one of them. At least one modem manufacturer offers a callback device which can store authorized call numbers, verify the list at every call received, then interrupt the communication and ring back.

The physical layer of a connection is well taken care of, but current solutions still leave much to be desired on the logical side. There is good news and bad news in this regard.

We will start with a system view, continue with the bad news, and conclude with the good news: what can be done to improve the current situation.

21.2   A SYSTEMS VIEW OF PC-TO-MAINFRAME COMMUNICATIONS

The five top goals we wish to achieve with PC-to-mainframe communications are: *link transparency*, including all aspects of logical communications links; *error free transmission; packet assembly/disassembly*, since the majority of attached devices work start/stop while mainframes should use packet switching; *data transparency*, whether we talk of text, image or data; and *automatic con-*

version: the PC works with the ASCII code while IBM mainframes work with EBCDIC.

Transparency and dependability should be complete not only in each current chore but also with a view into future requirements. The rational use of mainframes will typically be:

1. *Large database engines.* More precisely, text, data, graphics, image and voice warehouses with 1 trillion byte memory.

2. *Large communications switches.*

The future role of mainframes will not be data processing. DP can be done much more efficiently and at lower cost at the PC level.

Precisely for that reason we need a properly established communications discipline. Where the mainframe excels in terms of cost-effectiveness is in databasing. The PC should use the link to import and export files.

Here exactly comes the challenge, and for good reason. The PC to mainframe link is not straightforward. The bad news is that three groups of problems characterize such links:

1. Application level—or philosophical

2. Software level—or logical

3. Problems resulting from physical characteristics

For instance, even the 3270 PC and the IBM mainframe do not match. There are other physical differences which will be discussed at the end of this section.

Let's start with discrepancies at the application level. A bad start happened five years ago with the so-called "gorilla computers." This occurred during the time the DP/MIS department was insensitive to the micro, and users found themselves obliged to shop around on their own.

This brought into the organization many independent sources of supply which are not perfectly coordinated as to whether or not they run on local area networks (LAN). Because of incompatibilities, the best way is to weed them out based on the years that have passed by, and therefore, their depreciation.

Some good advice to follow on this matter is keep your PC as compatible as possible to the other PC in the organization and to the mainframes you are using. Ask yourself "Will the vendor be in business 10 years hence to support the installation and help in system solutions?"

This system solution should lead to the choice of the *system architecture*. The PC, supermicro, mini, midi, maxi, and mainframes in the organization should be networked. They should all belong to *one logical network*, though past practice may see to it that there is more than one physical network.*

---

*I have in mind a large money center bank which has 74 physical networks, including voice— 53 of them in the United States.

If you are a large organization, you should be choosing your global system architecture this year or at the latest, next year (1986). If you are a small organization, you can delay your decision for another couple of years, but remember, the more you delay, the more painful the decision will become.

Within the system architecture you will be choosing, PC and mainframes will interconnect. In choosing a valid system architecture, you would be wise to exercise much pressure on the vendor. He needs it.

The future of the network you select will depend on:

1. The money the vendor makes because of many adoptions

2. The pressure the clients exercise to get system development in full swing

The more the vendor is working on the interconnection problems, the better able the user will be in finding able solutions. Yet, not everything will be done on the vendor's behalf.

Able approaches invariably involve solutions to organizational issues. This is an internal problem and involves:

- users

- decision makers on information systems policy

- development responsibilities

It is time DP/MIS confronts its duties for the rest of this decade and for the 1990s. These duties are no more the blind writing of Cobol code. As Professor Wirth aptly suggested: "Writing Cobol today is a criminal offense."

Whether or not DP/MIS recognizes it, DP/MIS has the overriding responsibility to tailor the application to the end user. This means employing Fourth Generation Languages (4GL). The user can do some value-added programming through spreadsheets. More importantly, DP/MIS can improve the productivity of its professional programmers and analysts quite a bit. No responsible computer manager should miss this opportunity. Just the same, able DP/MIS management will be wise to steadily evaluate work in progress, chopping off the dead wood.

Whole forests can be affected by this accelerating sort of illness; *galloping obsolescence in DP*. After a soul-searching examination, a large New York organization established that 80% of the ongoing software projects (in development) had no reason for existence.

The able organization will not only promote man-to-machine and machine-to-man communications, but will also optimize in terms of used software. This means commodity software and 4GL will shrink to a minimum development time-table and bring the responsibility where it belongs: at the end user's side.

Fourth generation languages are easy to install. The problem is one of choice, among a myriad of alternatives. Another challenge is the need to build a controlled environment, so that the 4GL will not spread out of control, swallowing resources or changing computer use patterns.

Decisions and choices cannot be delayed. Fourth generation software technology is now entering the mainstream of computing. In many organizations, this shift is being fueled by advances in functionality and architecture and is bringing with it new interest in methodologies for applying technology most effectively.

Fourth generation tools provide application programmers with productivity improvements that are at least one order of magnitude higher than third generation languages. They are also accessible for prototyping purposes in cooperation with the end user.

A major asset of 4GL is that they can be easily learned and remembered. Another plus is work concerning software development can be done completely online.

The fast reader of this text may wonder what the differences being made have to do with PC-to-mainframe communications. The careful reader will, however, appreciate that the two subjects are linked intimately.

First of all, we need a systems perspective. Second, within this system-wide view we must define the applications. PC-to-mainframe connections are not done for pleasure, neither are they an intellectual pastime. They are made to meet requirements.

Such requirements must be in accord with the needs of end users, well established in technical terms, and properly integrated into a computers and communications aggregate. Typically, PC will be used for one or more of the following reasons:

1. Timely answer to user requests

2. Better response time—typically 1 or 2 seconds

3. Effectively distributed data processing and datacom operations

4. Greater machine intelligence at the workplace

The microprocessor under your desk replaces the dumb terminal, thereby improving response time and enabling more things to get done.

One doesn't care anymore if he/she monopolizes the processor—provided the PC which was chosen serves you well as the one and only user. Also, you can add on functions as you gain experience.

Because processing is now distributed on many dedicated engines, adding applications does not upset those already running. Implementation timetables can be significantly accelerated. But there is a technical problem in the PC-to-mainframe connection and it involves the type of processing.

• The mainframe has always been batch oriented

• The PC is designed as an interactive engine

These two design philosophies do not quite match. We can correct such a deficiency through software, hence, in an application sense, the need for a system architecture on which due attention has been placed.

The object should be to support interactivity in an environment in which mainframes and PC come together. Solutions should include all layers of database elements. They should be tuned to enhance the goal of system integration.

21.3 THE BAD NEWS: LOGICAL AND PHYSICAL DIFFERENCES

We said the applications-wise PC and mainframes differ. To appreciate how complicated the problem of linking them together is, we should look at their *logical differences*.

The first important reference is that of *character sets*. Mainframes (particularly IBM computers and compatibles) work in EBCDIC.* PC work in ASCII—and ASCII comes in 7-bit and 8-bit codes, plus various dialects.

The knowledgeable reader may say: "No problem. We can do code conversion." That's precisely what we need to do, but we should also remember that conversions cost money, involve time delays, and often are at the origin of errors.

Such a code difference is unfortunate. An elegant solution would have been to map the microfiles (personal databases) on the PC onto the mainframes, without undue interfacing. Properly used, PC offer important facilities to the user (and to the organization) who can employ them in an effective and coordinated manner. The integration of micros into your current DP environment should be not only a goal, but a continuing process.

Therefore, it is important to identify incompatibilities so that we can take positive steps to build a foundation for continuous integration, able to extend into our future systems planning and development activities. The second reference on incompatibilities is the *instruction set*.

The instruction set of a mainframe typically features 132 to 200 instructions; that of a PC less than 96 instructions.

The difference is not just one of numbers. For every practical purpose, these are *incompatible* sets. Claims that you can port programs from mainframes to PC are largely unfounded. As we will see in the following paragraphs, there are other reasons compounding the forementioned problem.

The third major difference is in the *higher level languages* that mainframes usually support: Cobol, Fortran, and PL/1. PCs work best in Pascal, "C", and Basic.

Even if the PC runs Cobol (which will be highly unadvisable) the Cobol dialect will not be compatible. There are an estimated 250 Cobol dialects after all. Even Cobol ANSI has a dozen not-so-compatible variations.

The fourth reference to incompatibility is *program size*. The PC has a small central memory and the hard disc is not always there. Sectioning programs and overlaying them is the most efficient procedure. It has been tried in the 1970s in the passage of software from mainframes to minis and failed. It is unwise to repeat a bad experience.

---

*Extended Binary Code Decimal

Fifth, there is a problem of *protocol translation*. The type of protocol presently used with mainframes is packet switching (SDLC) or bisynchronous (BSC). PC usually work start/stop.

Through packet assembly/disassembly (PAD), start/stop devices can effectively communicate on a packet switching network. But here again we talk of translation with all the negatives to which reference has already been made.

All this is not written to discourage PC-to-mainframe communications but to enhance them. The aim of bringing under perspective prevailing incompatibilities is to identify that integrating microcomputers into a mainframe environment is not limited to the simple attachment of a PC to the mainframe as a terminal device.

While the physical link is a requirement, it falls short of true integration. Over the last five to seven years PC users may not have planned for direct access to the mainframe, but they want it now. The foregoing discussion which helps document this task is more complex than a first glance reveals. For instance, a file transfer system that couples microcomputers to existing applications and data structures is critical to this effort. Indeed, a file transfer system is the first step in integrating micros into a data processing environment.

The sixth reference is *proper data handling*. This starts with database organization and, as well, database access.

• Mainframes have complex database structures—and there is quite often a labyrinth of solutions regarding this reference, never streamlined as they should have been.

• Micros have smaller files which are more efficiently organized, often running since the beginning under a database management system (DBMS).

PC files are often sequential; sometimes index sequential. With mainframes, ISAM and VSAM are the rule.

One application at a time runs on the PC. Mainframes have foreground and several background jobs. The mainframe operations are characteristically: multiprocessing, multiaccess, and multicomputing.

The speed of the central processor conditions the choice—which is also influenced by past practice to a very substantial extent.

To simplify the solution we are looking for, the concept of information centers has come around. They assure that mainframe access is simple, effective, and fast from the end user's viewpoint. They also verify that data security and integrity controls exist.

But while serving the end users, an information center implementation greatly supports the DP/MIS director. As we will see, it helps assure that there will be no mandatory application changes, nor that the DP will be forced into major application stream overhaul.

The seventh and last reference on logical difference is that of basic software. It ranges from the operating system to the DBMS and the transactional routines.

Leaving aside Unix (particularly Version 4.3 BSD) which has networking characteristics, no operating system today effectively supports a networked environment. But mainframes have featured, for over 20 years, frontend processors which interface between the classical OS and the network. This is not so with PC. However, when attached on a local area network, micros connect to the cable through bus interface units (BIU) which play the role of frontends. Such communications-oriented layers do not exist with long-haul connections.

At the rearend of the computer, both mainframes and PC use DBMS. The latter are mainly hierarchical or of an owner/member (networking) structure. Relational DBMS only recently appeared for mainframes. DBMS for micros are simpler, more elegant and usually of a relational nature.

Let's now look into the *physical incompatibility*. To exchange information, PC and mainframes need input/output compatibility—which practically means common forms. This is not necessary with the two types of engines we are considering.

First, in terms of *input*, mainframes originally worked batch, getting their input from punched cards and magnetic tapes. The PC has neither punched cards nor magnetic tape. It is keyboard based, may have a graphic tablet, touch-sensitive screen, mouse or similar cursor positioning, and/or input device.

Second, regarding *output*, mainframes have classically been hardcopy (HC) oriented. PC are designed for softcopy (video) interactivity.

Third, the *physical connection* is less than perfect. It could be done in an *indirect* manner, not online, porting a floppy from PC to mainframe—but that is a solution which is better to leave out, except for backup.

Storage on floppy disc and floppy disc exchanges is messy, but in case one wants to apply this solution (even if irrational) he will be in for some surprises.

- Mainframes typically have 8″ floppies, rarely 5¼″.

- PC most often use 5¼″ floppies, and the new standard is less than 4″ (3¼, 3½, 3¾ inch diameter.)

While the latter example suggests a physical incompatibility, chances are great the 5¼″ will be logically incompatible. Even within the product line of the same manufacturer, one machine cannot read the 5¼″ of the other. It happened with the Tower (supermicro) and DM V (PC) products of NCR.

Fourth, chances are PC and mainframes will exhibit *modem* incompatibility. Mainframes have sophisticated high-speed modems working at 19.2, 56, and 64 KBPS. As stated in the introduction, PC work at 1.2 KBPS—and most usually at 300 or 600 baud through acoustic coupler.

Fundamental modem differences lead to an incompatibility in *lines*. Here the reference is similar to the one being made about modems—as well as to what has been said in the discussion on logical differences about protocols.

The whole issue revolves about the philosophy of the application. While, as it has already been discussed, the PC can work as a dumb terminal, basically this is not a good approach in the long run.

The PC should not be, physically nor logically, hardwired to the mainframe. Let's recall that in the physical connection, the non-intelligent cable is 30 meters or less—while baseband LAN goes beyond 300 meters and with broadband LAN in the 3 to 10 kilometers range.

A baseband LAN will typically be digital, may feature time division multiplexing (TDM), and work in the 300 KBPS to 10 MBPS range. The protocol will tend to be carrier sensing multiple access with collision detection (CSMA/CD); it will feature no priorities, and use twisted wire, flat cable, or flexible coaxial cable.

A broadband LAN will tend to be analog, divide its bandwidth through division multiplexing (FDM), and feature 400 megahertz (MHz) (for protected coaxial) or up to 1 gigahertz (GHz) capacity with optical fibers. It will most likely work with token passing, providing priorities.

21.4   THE INFORMATION CENTER STRATEGY

Knowledge is power. Our post-industrial society can operate only by having information open to everybody and by getting each individual to respond. Our future depends on shared information.

Knowledge and information have two special characteristics contrasting them to physical assets. First, the more we share them, the more of them we have. Second, if we stop sharing them, they decay and eventually become negative assets.

The definition of an *information center* (ICenter) fits squarely into these notions. It is built to serve three basic goals:

1. Help share information among top management, middle management, and professionals in the organization—specifically the people who have been the least served during the first 30 years of the information revolution. The Information Center does not address itself to transactional activities. These are more or less well supported by the classical realtime operations. The goal of the Information Center is management service. This is consistent with the fact that to gain competitive advantages, an organization must have steady and easy access to timely and accurate information. This means that information center design must be based on understanding of this technology, to deliver it properly to users.

2. Dissociate the management information system activities from the transactional chores, thus avoiding a confusion of the chores by transforming the current day-to-day activities into a two-tier system. Not only have classical DP center operations been basically transactional (money-in and money-out) oriented, but they also have had a twisted history of development. Files are rarely integrated into a coherent, comprehensive database. Just the same, transactional procedures do not necessarily use DBMS—some of them do and others don't.

With two to ten million Cobol statements in the library, it is unwise to move too much in the foggy bottom of classical DP activities. The risk is there of destabilizing the system. Hence, the information center approach: Linking the new management oriented database, which is typically relational, to the old one (usually hierarchical) through data extract routines. This provides significant freedom in design and leads to the third goal.

3. Using the expertise we gained through 30 years of computer practice and 20 years of communications experience to develop a comprehensive information system for managers and professionals. Such a system will be typically supported by mainframes at the level of the *text and data warehouse*, and by PC under the user's desk.

What we are essentially developing is a rich and actual internal database, an able complement to the long list of external databases the manager and professional user of the information system is welcome to access. First, this involves a good deal of *planning*, that is, defining the ICenter and identifying who will be involved and how.

The second basic prerequisite is *designing*. It includes selecting system architecture, OS, DBMS, equipment and staff. It also includes defining the policies and overall strategy.

The third key issue is *developing* the stage at which products and services are chosen. This, too, is instrumental in putting the ICenter into practice.

Fourth, there is need for properly managing and evaluating the time at which plans are put in action. There is also a need for ICenter activity effectiveness evaluation.

Both planning and designing must be given high priority because we now have the experience to do the work in the correct manner. Indeed, an estimated two-thirds of an information center's troubles can be traced to mistakes made during the start-up phase.

The right planning must cover technical issues:

- initial information (product) selection
- possibilities of networking PC
- the maintenance of information elements

Better results are obtained where there is corporate management involvement, and when the concept of making the ICenter a valid service organization is present.

End users should definitely be involved in planning the ICenter, promoting teamwork and building long-term commitment to technological solutions. At the same time information systems policies should be important in improving senior management support.

An information center implementation will be successful if company information is re-sorted and custom designed for the executives. The ICenter should allow them to track key data and strategic statistics through their PC in the center's database.

Here, the user's collaboration is most important. In all likelihood, failures will be due to the lack of strategic direction, no clear plan of action, and a deficiency in the means of achieving the ICenter's goals.

Once the nature and type of the information elements to be included in the ICenter have been properly defined, technical issues come into perspective, for instance, data architecture and the process of developing a blueprint for the organization's information structure.

Technical issues invariably involve structured systems development, that is, a methodology which provides the building aspects. There is also need for data administration: the process of managing the information resource; and for effectively linking the distributed PC to the ICenter mainframe.

Information center decisions can be far-reaching, and simple mainframe-to-PC file transfer plans would not anwer requirements. The solution to be adopted should cover all requirements for what information users consider important.

This brings PC-to-mainframe communications under a new perspective. It addresses both internal ICenter resources and public databases:

1. Public data are volatile, voluminous, and used at a deep level of detail.

2. ICenter provides strategic information derived from transaction data and used by managers and professionals to analyze and evaluate their business activities.

3. ICenter data is time-dependent, aggregated, low volume, and represents quantitative status of a business.

4. The user's WS is where data manipulation takes place, with analyses performed to support business decisions and to document them.

This outline of key functions highlights critical success factors for the information center. We have spoken of a number of variables: planning, designing for service, information element definition, system development, asset protection, and application strategies. Success must be defined within the specific context of each organization's culture, but general guidelines can serve.

21.5 GOOD NEWS: THE SOLUTION IS A SYSTEM ARCHITECTURE

We have said that the simple link of PC to the mainframe is far from reaching the goals an enterprise should set for itself in computers and communications. Much more is necessary, starting with a far reaching information system strategy.

The choice of a system architecture and the development of an information center policy are two of the pillars of this strategy. Another basic reference is finding added value application for the PC to be installed at managerial and professional workplaces and also finding and implementing the appropriate commodity software.

Just as important is training the users in computer literacy. This requires hands-on experience with spreadsheets, graphics, some word processing, and quite definitely electronic mail. The last five years of PC and mainframe experience help document that actually what users want to do with computers is to communicate rather than to compute.

If we look back to the first application of electricity, we see that this was the *telegraph* (1834), a communications service. Electric light, the Edison invention, came twenty years later.

Just the same, the *telephone* was invented in 1873 while the typewriter took another 10 years, and calculating machines of the punched card type appeared in the early twentieth century. Throughout these developments, the communications drive was foremost.

As a product, the telegraph was followed by *telex* in 1946, and by *electronic mail* three decades later. Today in business and industry, an electronic mail service offers the best means of information exchange with quick filing and retrieval of messages.

Commodity software helps organize messages by sender, receiver, and interest group so all messages on the same topic can be handled together. Users can selectively read mail, store and review messages, join a conference already in session (in computer conferencing applications), and read all past messages associated with that conference.

Personal archive files store messages and serve as a postdate reminder. They date messages that do not require immediate attention, and software is reminding the user when the dates come due.

Electronic mail and computer conferencing systems are user friendly, typically working through menus. They feature full-screen access with easy-to-understand prompts and help messages. They transmit files created with commercially available software to other PC and mainframes.

In fact, electronic mail is one of the better examples of PC-to-mainframe communications. It also constitutes a good documentation of the fact that it makes no sense to have the PC working alone. Interconnection should be the order of the day.

The point though is that interconnection underlines the need for a *system architecture*, under which will integrate the information resources of the organization, from mainframes to PC.

Many subjects should be considered and supported within the network architecture to be chosen: data security, passwords, encryption, authentication, authorization, data integrity, recovery, restart, logging, and journaling. But the key contribution of a system architecture is that of *a network-wide frame of reference* within which integration of discrete, and often diverse, software and hardware components will take place.

In this sense, the PC-to-mainframe connectivity should be limited to the data link. The work to be done should be directed to the compound electronic document. This is a goal well served by the system architecture.

Benefits for the system architecture are multiple and profound. It permits a *stable* basis for analysis and programming. It gives a common direction on development.

The system architecture not only helps integrate software and hardware belonging to diverse services, but also it assures that technology will work for the organization applying the architecture. That's why the choice is so important, and often difficult.

A different way of making this statement is that the PC-to-mainframe connection must have goals. These are your applications. The applications should drive the system—and the architecture should be serving them.

A good system architecture will invariably change some of the computers and communications landscape. Classical mainframe systems have been created with one important assumption: The machine that stores the data handles all the processing of the organization's data. This is no longer true.

Processing should be distributed in well-defined ways. Query languages ease the user's access to databases. Electronic mail links users (and their PC) to one another.

Today we face the prospect of delivering data to the end user for processing on a computer that is not under central control. This introduces the problem of multiple solutions which don't fit together, hence, the need for architecture *even if* for no other reason than creating a common denominator.

Now is the time to take a very thoughtful look at system integration. We must carefully examine the evolution in computers and communications, study the systems aspects, establish the information center responsibilities, and link all logical and physical devices at our disposition into *one aggregate.*

Chapter 22

USING THE NETWORK

22.1 INTRODUCTION

The term *computer network* is currently used to describe a wide range of data processing, databasing, and data communications facilities. In the most general sense, a computer network may be seen as a collection of computers and terminals, linked via a communications system. However, employing a single term to describe such a large variety of functions masks an important distinction that exists among networks:

1. The case in which *the user must explicitly manage* the computer resources.

2. The alternative case, when these *resources are managed* automatically by a network operating system, and are transparent to the user.

In a way, this is a repetition of the experience which we have had with computers: thirty years ago, it was the human operator who ran the machine, but by the early 60's the evolving complex computer equipment was run automatically by the *operating system*, which consumed a lot of power in the process.

In the class 1 of networks, the responsibility for resource administration falls upon the user, whereas in class 2 the user can depend upon the aid of a network operating system in the acquisition and handling of needed resources. This distinction is more appropriate with reference to distributed rather than to centralized computer networks.

The aforementioned distinction between class 1 and class 2 is based upon the different ways in which a user may view a given network, according to the degree of transparency presented by the network to the user. Networks in each of the two classes share an important characteristic: the operation of the network and the management of the communications systems tend to merge.

The single computer system linked to remote terminals by means of communications facilities was the earliest system generally described as a computer network. Such centralized systems are still very common today. Nevertheless, a processing facility which includes more than one host* computer is not necessarily a computer network. (Excluded from the concept of a computer network is the conventional multiprocessor system, in which processors are linked through their use of shared memory.)

To qualify as a network, host computers must exchange information by means of communication lines: whether local or long haul. We should note that, according to this definition, the host computers of a distributed network need not be geographically dispersed; in fact, in the extreme case such a network could be contained in one room alone. The basic characteristic of a computer communications network then is that the user views the network as a collection of several computing systems with varying services and capabilities. From the many computer systems available on the network, the user must explicitly choose the one best suited to run his job. Choices are called for as many varied resources are offered to the user, including gateways to other networks, personal local and remote databases, applications programs, specialized hardware units, and so on.

22.2 ROUTING, MONITORING, JOURNALING

To access the computers and communications resources, the user of a computer network must determine the systems on which the resource resides, and familiarize himself with the commands necessary to invoke a given resource, on a particular system, at a particular time. But if there is an overall operating system for the network, the user may not even be aware of which processor handles this job. The choice of the processor to execute a job is transparent to him.

When the resources are automatically managed, the network operating system makes decisions for the user. In the alternate case, the decisions are made by the user himself. In years past, the user who wished to access different computer systems on a network had to be conversant with each system. This is no more necessary as this job gets automated through the provision of a common user interface run by the network architecture.

Another very important job the network architecture is doing for the end user is managing its own frame of operation. This particularly regards three functions:

1. Routing the messages

2. Monitoring to assure that everything happens at the right time, in the right order, and

3. Journaling—keeping track of all activities transiting on the network

The data flow between the switching computers will be typically controlled by ordinary line control procedures. Line management can involve special soft-

---

*A computer where user programs are executed.

ware in the computers on which these processes are running. Routing concerns itself with the channeling of this data flow.

Modern data communications networks use software control programs residing at host computers to control interprocess communication. This can be done if all of the computers and systems are similar; if a frontend device performs housekeeping functions between the network and the software controller; or if the same protocols are being observed creating similar logical engines out of the dissimilar hardware machines.

In this set-up, the nodes switches are acting as controllers responsible for flow control, accepting inputs from all processes in remote computers (microcomputers, minicomputers, mainframes, terminals) desiring to communicate. The network's backbone will be composed of:

1. Nodes communicating with other nodes, hosts, terminals, concentrators

2. Lines over which messages and transactions are being transferred

The network's nodes act as switches. Their node-to-node communication can be functionally partitioned into several distinct functions:

- line interface
- buffer management
- process interface
- supervisory activities.

A supervisory record can be 48 bits long (Tymnet). It travels along the data route. This exchange of supervisory records is transparent to the user. The main functions of the information network are presented in a comprehensive form in Figure 22.1.

Routing is an integral part of each network which presents alternative paths to the transmission of data. Messages are transmitted in a given form through a given path. When a message can transverse more than one path in a network, from source to destination, then the communication software must perform this routing function.

Routing overhead comes from three sources:

1. Nodes report any links that are out of order by sending a supervisory record every so many seconds.

2. Supervisory records instruct the receiving node(s) to change the "Permuter Tables"—and such instructions must be executed.

3. The ACK of these changes makes them effective in a supervisory sense.

Such operations take place without any particular concern to the user—whose primary interest is with his system rather than the network architecture. This is

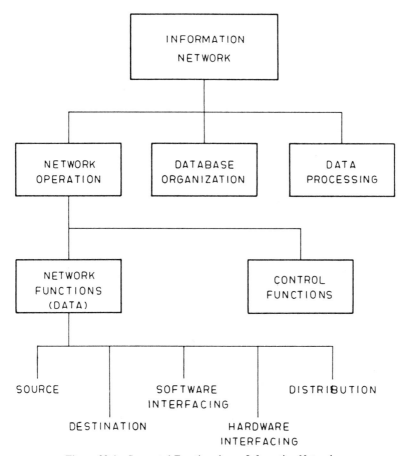

Figure 22.1 Supported Functions by an Information Network.

the basic benefit from which derives the interest to the services of valued added networks (as we will discuss in the appropriate section).

Let's see how routing is implemented. If the network is circuit switched*, then a physical connection is established at each switching point, and remains active for the duration of the connection. The software involvement in this case is minimal, and can, for example, control an automatic calling unit.

If a logical connection is established, then the communication software residing in some network component (or components) must select the best path for the establishment of the circuit. This can be done by comparing the conditions of the possible physical circuits to select those with some combination of light traffic and least errors.

---

*The establishment of a physical circuit between nodes prior to the start of transmission.

• *Fixed Routing* involves one and only one route between two stations in a network.

• *Dynamic Routing* presupposes that the data communications network chooses a route on its own from messages to destination.

This is, by the way, what happens with the circuit switching telephone service.

In a message switched network, each individual message can be routed dynamically or messages can be subdivided into packets, each of which is dynamically routed. The objective is to have the circuits efficiently utilized and to obtain a high throughput. The same is valid with packet switching.

A vital issue to guarantee is *Information Integrity*. This requires that the network components maintain the specified (or required) accuracy of the information flow: Parity, Block Check, Hash Totals, Cyclic Redundancy—of which we have already spoken—are means to this end.

Restart and Recovery procedures are part of the process designed to guarantee information integrity. And so is Journaling. *Journaling* is the ability to retain in one or more storage media (e.g., tapes, discs) copies of information that has passed through the network, and/or was delivered to the intended destination.

Networks can include Journaling capability in one or more locations. However, the maintenance of journals presents both logical and physical problems—and, also, issues of security.

Another vital function is monitoring. *Monitoring* can be achieved both through hardware and software, distributed throughout the network. Actual monitoring is usually accomplished from one or a few control centers. Figure 22.2 demonstrates the basic control function in a network.

By intelligent use of automatic remote checkout logic, including remote sensing of component failure, degradation and shut-down can be detected and personnel dispatched for either node or channel repair. Degradation is a gradual deterioration in performance. We can generate statistics to measure the internal performance of the communications network. For instance, queue lengths, message sizes, line loads, time-up, overall response, routing, to aid in tuning the system and providing fault detection and correction.

Statistics on traffic flow between nodes must be analysed to determine proper distribution of facilities, including location and number of host computers, switching computers, device interfaces, bandwidth, and layout of communication circuits. Usage measurements can be made of the utilizations of the different processes available through host computers and of the types of services offered such as interactive, batch, or graphics.

Both the internal and external measurements will aid in determining the effectiveness of the computer network, the appropriate distribution of services, the need for expansion (or contraction), generally providing information for management decision. These subjects together make up the key topics of an effective network management activity.

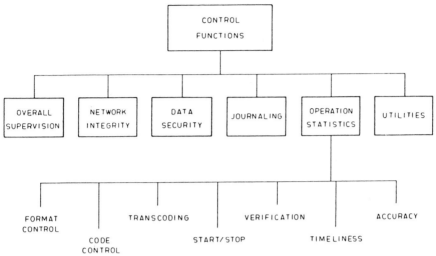

Figure 22.2   Necessary Range of Control Functions.

22.3   NETWORK OPTIMIZATION AND UTILIZATION

The terminals, geographically distributed, operating with their own characteristics—and through similar or different types of line disciplines—will require both hardware and software resources and interfaces to function. With communications systems we define two fundamental ways to line connection:

- point-to-point, and
- party line or multidrop

With a party line is multipoint. Each terminal is connected to the principal line (backbone), which initiates at the mainframe or minicomputer, through shorter line: the derivatives. A party line configuration must use dedicated (private) lines and the appropriate protocols. We mentioned polling-selecting and hub polling as examples.

In the preceding chapter it was also said that to improve the performance of a private line (and, as well, of the public line) we may proceed by concentration. Still an earlier chapter made the reference that data transmission can be effected in parallel or in series.

Such issues bring into evidence that communications lines can be optimized in configuration and cost by effecting the appropriate choices. The problem of optimization must be faced within the realm of the utilization to which the network will be put. For instance, the choice of line speed or of a particular technique of concentration will be meaningless without first specifying:

- the use of equipment
- the data load
- the needed response time
- the allowed costs

It should be recalled that the network applications vary widely in the ease with which they may be implemented and the benefits which they will yield. Some relatively simple applications may offer paybacks, but some others, which will require greater effort and longer periods of time to effectively implement, may provide even greater benefits.

The crucial factors to keep in mind in evaluating the performance characteristics of a computer network are a fairly simple list: Both hardware and software aspects are involved in these characteristics, but by and large, the criteria used for this or that choice should be applications-oriented.

1. The most obvious use of a network is for the wide and relatively inexpensive access to resources not available locally. Such resources may include "new" computer power on a variety of machines, software, and databases.

2. Networks make possible a reduction in the cost of accessing these resources by permitting users to share communications facilities. But costs go down as utilization of the lines goes up.

About the only applications which would be very awkward to attempt through a network are those requiring manual intervention through the main computer console. Few programs are written this way anymore. Most network architectures provide for *pass-through* capabilities operating the attached computer resources on a network through a central console.

3. Standardization and remote access help both in cost reduction and in backup facilities.

A network can provide backup for a local installation, if some advance precautions are taken, such as locating duplicate copies of essential files at sites in the network. The same principle applies to the regular use of the components on the network. By storing duplicate copies of files at other locations, the failure of a particular node will not prevent essential work from being done elsewhere.

4. The access and backup benefits are derived from the ability of one user to access many resources.

5. A primary benefit also derives from the ability of a single resource to serve many users.

For resources with high fixed costs but moderate variable costs (such as the costs associated with developing and using a software package) benefits are realized when utilization increases. This results in lower unit cost for all users, and may make feasible specialization which could not otherwise be cost-justified.

The orientation of a data communications network toward the broad market of the personal computer users is a good example of putting into effect the advantages described in the last paragraph. One way of looking at the fact that 95% of all American homes have a telephone is to a vast potential population of users, far outpacing business and industrial needs in data communications.

The user of a computer network is a customer for computers and communications services. He is unconcerned with the technical details of how these services are produced or how they are delivered to him. He usually becomes aware of underlying technical factors, only when they somewhat fail to meet his requirements.

The most serious technical problem facing computer networks is to provide service with sufficiently high reliability—at reasonably low cost—that customers may safely come to rely on it. Reliability standards must be strictly upkept, including the ability to rapidly diagnose and respond to failures in the network.

Network control centers serve the functions of continually monitoring problem areas, coordinating corrective measures and providing a central point to which users may direct inquiries and complaints. Stated in different terms: When a computer is so employed, the user is unconcerned with how the services are produced. Concern and judgment are directed to the *end product*, the information service:

- How well does it fulfill expectations?
- How correct is it?
- How promptly is it delivered?
- What's its cost-effectiveness?
- What is its quality?

Flexibility is yet another consideration. Networks, as they presently exist, still do not permit all sorts of devices to be connected to them; or all interconnections to be made, which may be desired. The problem is both hardware and systems software. But there is also the question of protocols.

Protocol problems are generally responsible for the inability to interconnect networks or to accomplish meaningful data exchanges between computers of different makes. In this sense, the standardization of the presentation level protocol (NA-PLPS in North America, CEPT in Europe), within the context of the ISO Open Systems Interconnection we treated in the chapter on Protocols, is good news.

Security is an issue which has been conveniently ignored by most networks, but which must be faced too. (Even in an academic environment, there may be found instances of sensitive files, which require adequate protection from unauthorized examination and tampering.) Networks which offer service commercially have the responsibility of developing measures to protect their customers' sensitive information.

Furthermore, standards are needed to permit both the access to multiple networks by users with a single set of terminal facilities, and the eventual interconnection of networks. The importance of standards to the user is that they facilitate the interchange of software, hardware, and people.

- *Software standards* permit programs to be run on many different computers.

- *Hardware standards* allow terminals and other peripherals to be connected to many different systems.

- *Procedural standards* make it easier to move from one network to another as their needs dictate.

Potential users of network services will want to inquire as to which standards the candidate network conforms. Experienced users and those with special requirements may wish to participate in future standardization efforts.

Economic problems (cost/effectiveness) are another key issue of data communications. Since network designers are most frequently computer specialists or communications engineers (and since the early computer networks have been largely experimental), the economic and financial problems have not been addressed as they should. Such problems, however, can be every bit as crippling to successful network operations as the technical ones.

Financial problems include the costs associated with the various network components, and the raising of the necessary capital to develop and operate a network. Computer networks are expensive to develop and operate, with substantial costs arising from both hardware and software components, as we will see in the following section. Standardization is a way of helping reduce costs, by making the hardware/software/protocol components interchangeable and, hence, available to the largest possible population of users.

Standardization not only enhances quality protection, but it also makes feasible a critical mass. Economy, quality, and standardization are subjects complementing one another. Furthermore, the worth of the Value Added Network (of which we will be talking in the following section) will be that much greater if protocols and services are standardized.

22.4 VALUE ADDED NETWORKS

As society has progressed from its primitive origins, the problems related to economics have become more and more complex. New products, new processes, and new systems have underlined this basic fact.

This is the case with computer networks. Not only are they expensive to develop and operate, but also most costs are fixed over a given operating period.

The result is a high sensitivity to variations in demand and usage. Hence, the need to raise network loading to the point where benefits are realized.

Another challenge to computer network developers is to reduce overhead and thereby the costs associated with such components as hosts and software specific to the network. The communications components, too, represent opportunities for savings resulting from economies of scale.

Let's recall that a network involves a myriad of components: hosts of all types, terminals, modems, lines, switches, control centers, and software—lots of it. Not only basic software but also application programs can be provided by the VAN, downloaded to the end user. Then there are services, a lot of control procedures, circuit setups, interfaces, and so on.

Every component in a network has a price. From the user's standpoint, the network costs are those he must sustain for interconnecting his own computer, minicomputers, microcomputers, and terminals through the network.

The costs include providing the actual interface for computers and terminals to the communications lines. This and the associated cost factors are derived from published rates for existing and proposed LAN—to be precise, computer communications networks that utilize the existing common carrier facilities for transmission, while providing added data service features with separate equipment.

(These added data service features are typically implemented with computers programmed to provide store-and-forward message switching, routing, journaling, terminal interfacing, error detection and correction, host computer interfacing, and so on, completing the interconnection of a host computer to terminals, and possibly, to other host computers.)

Judging from the services which it offers, VAN is a communications facility and provides no end, or information-based service. This definition does not necessarily cover the entire range of situations, to which the term "value added network" (or carrier) has been applied; nor does it address legal or regulatory considerations. But it does exclude the trivial cases of networks from its coverage and addresses a new type of data communications service where advantage is taken of improved terminology.

Examples of existing and proposed VAN covered by this definition are Tymnet and Telenet in the U.S.; Datapac (Canada); EPSS (U.K.); Transpac (France); Euronet (Common Market). And the USA, Canada, England, and France are the first countries to launch themselves into value added networks through private or public effort. Germany and Japan are other examples.

(It is, therefore, no wonder that these countries played the leading role in the definition of the new standard for "Interface between Data Terminal Equipment" [DTE] and "Data Circuit Terminating Equipment" [DCE] for terminals operating in the Packet mode on Public Data Networks—known as "Recommendation X.25".)

The discussion on value added networks necessarily brings attention to the structure of a distributed information system and the way it works. Since Part One we have underlined the wisdom of distinguishing four major issues:

- distributed processing
- distributed databases
- distributed datacomm
- distributed end user functions

also appreciating the fact that there is an increased user's responsibility in terms of quality of input, sensitivity to output, and timely response to his own information needs. There are other issues to be appreciated.

A network which supports computer-to-computer communications (as opposed to just terminal-to-computer communications) may permit programs to be written which operate on more than one computer system simultaneously. This is distributed processing through downline loading, where different parts of the same program may run simultaneously on several central processing units of a single computer system.

In fact, not only processing but also databases can be distributed and downloaded to fulfill the aim of bringing the computer power nearer to the user. A distributed database is an approach wherein the user's files are partitioned or replicated into parts. Each part is assigned to different information processing nodes of a distributed system.

A value added network does not only transit text and data for computer equipment the way we usually look at them. Facsimile transmission is a different example. It brings fully automatic text, data, and image services to every office in a format designed for convenience.

Modern facsimile units offer autodial, auto feed, auto poll, auto answer, and status reporting at CCITT Group I, Group II, Group III and Group IV speeds. This means they communicate with any

- Group I Analog (North American EIA, 6 minute)
- Group II Analog (3 minute)
- Group III Digital (sub-minute unit)
- Group IV Digital (4 to 6 seconds)

Versatile facsimile transceivers produce high-resolution copies in just a few seconds, and can reduce oversize documents to a standard letter size during transmission.

Detailed graphics such as engineering drawings or intricate forms are reproduced with good legibility. And such devices are usually designed for use by anyone in the office. Their ease of operation makes them valuable to a business environment.

Distributed solutions help increase the user's responsibility in handling his own workload. As such they are a reversal of the trend toward centralization in information processing which characterized the 1950s and 1960s—but is still around in some quarters.

There was a time when state-of-the-art limitations forced users to place all their computer resources at a distant central site and then to adjust their business operations to meet the restrictions imposed by such centralization. But *today* technology makes feasible the distribution of computer power (processing, database, datacomm) in a way that best fits the user's needs—with only as much centralization as is required.

The aim is to locate the processing and storage capabilities close to the workplace and to give the man with the problem the possibility of solving his problem online. Distributed systems help:

- assure interactive man-machine communication

- shorten the input and reporting route

- eliminate the delay inherent in round trips

- provide new services, electronic mail, and videotex being examples

This way we avoid repetitive clerical operations, provide for error detection and correction online, obtain better response time, reduce the overall cost in data handling, simplify programming tasks, handle more transactions in the process, and automate the fringes of the operations in the periphery—which so far defied computers to accomplish such goals. Jobs can be transferred between computer resources—from personal computers to mainframes, the data load can be leveled; the failure of one center will *not* affect the whole system.

Such a structure typically supports data entry operations where the information originates by providing the appropriate computer power where needed, assisting in editing for errors, and forwarding the correct messages to the right destination. A network-based distributed environment puts more capability where it is most necessary: at the remote site. In the last analysis, it benefits the ultimate user, as the work is done near to his site, in the proximity of the system's "ends."

22.5 VIDEOTEX, ELECTRONIC MAIL, HOME BANKING

Videotex systems are communications engines for text, data, and graphics—hence with the ability of pictorial representation and a repertoire of display attributes. The text and the pictures obtained are intended to be displayed using the current television (TV) standards of the different countries.

It is a recognized way of approaching user access to databases so that end user actions are intuitively obvious. No training or manuals are needed. Whatever the user is doing, the available operations at the terminal are limited and consistent so he can not get lost or crash the system.

Videotex can be seen as a frontend to the database, placing an envelope around the traditional computer user interface so that the actions available are limited to:

- Selecting an item, from a menu
- Moving to the next page in sequence
- Backing up one page
- Backing up to a previous menu

Throughout the system, these actions are always done the same way. Information appears a page at a time. This information is enhanced with color and graphics to make it more meaningful or easier to read.

The restrictions in protocol choices do make videotex inappropriate for applications requiring many functions or complex interactions. For some applications the potential performance impact of the graphic, color, and formatting components is offset by the limitation of page-at-a-time access.

However, the overall approachability and obviousness of the interface extend the usefulness of the database or transaction processor from the few knowledgeable computer users to the millions of people in homes and offices who have never used a computer. The videotex system will typically include or access private databases, providing information access to subscribers.

Gateways are the means by which existing computer databases can be linked to packet switched networks.

Information Providers (IP) are in the business of supplying and selling the information included in these databases.

The ability to access information on product availability, performance or pricing is a benefit the retail and wholesale trade will put to real advantage. With videotex technology, this can be a reality for even the smallest shop. It can help suppliers and manufacturers build up the information services they need to promote products and keep both customers and salespeople up to date.

The use of videotex for collecting information also indicates important business applications through its convergence with data processing. For instance, sales orders can be typed into a videotex system and processed by a connected mainframe. Also, information held in a mainframe computer can be accessed by untrained staff through videotex, thus helping to reduce the alienation caused by the introduction of computers in business and industry.

Videotex also has application in training. In England, one major bank has installed videotex to train its employees on banking procedures. These are stored on a computer and can be accessed in any of the bank's branches, replacing traditional classroom training methods.

These same Vtx features provide support for home banking (HB). In the organization of the HB services, a study identified four layers of interest in the implementation and public distribution of a videotex system:

1. system operator
2. gateway support
3. information provider
4. end user

A basic question which has been debated relates to the type of offering: "Is a bank the right organization for system operator?" Answers are divided. AT&T says *no*—but Chemical Bank says *yes*, and offers its *Pronto* system both to its clients and to other banks willing to pay a fee.

Though the issue is far from being resolved, several financial institutions seem to lean toward the idea that a bank can be a limited systems operator. It should not undertake the full functions—even if systems like Pronto expect to put on other services such as merchandizing.

Delivery of home banking is a complex data processing, databasing, and computer switching problem. Its success depends on a number of factors, the most important being:

- the way it is marketed
- the range of supported services
- tariffs and pricing costs
- the easiness with which the end user can communicate with the system

At the beginning of this era of online financial services reaching the consumer site, terminals must be simple and low cost. That is what videotex practically does. At the same time, challenged by non-banking institutions the banking industry has to come up with solutions.

The danger videotex poses for banks is that non-bank financial institutions, which have not invested in brick-and-mortar branches, can reach the retail customer with a competitive level of product offering. A brokerage firm would offer its asset management account, which can usually be accessed by check or debit card, through a home terminal.

Merchandizing companies can do the same for their wares—expanding in credit handling and banking functions. The second largest financial institution of Atlanta, GA., is Rich/Richways—and that's a merchandizing firm.

These examples help identify that several factors are expediting the rapid growth of videotex. Technological developments continue to make it more widely affordable and expanded transmission media are increasingly available. Industry push sees to it that sectors such as banking, broadcasting, cablecasting, publishing, retailing, travel, and transportation promote videotex commercially as it offers them the opportunity to extend or defend their markets.

Consumer acceptance has been overwhelming in the different tests. A series of controlled tests that date back to 1978 has determined that high quality videotex services stimulate consumer demand. Other critical factors are the growing emphasis on computer literacy in schools and familiarity with the increasing number of interactive devices in the office.

The ready acceptance and popularity of basic interactive devices, such as remote games, is changing the home screen from a passive to an active medium; consumer interaction with thousands of automated teller machines helps promote videotex usage. The same is true of the rapid growth of personal computers.

Personal computers and videotex are no competitors to one another, they are complementary. Currently available capabilities are in the middle of a long line of developments which have started last decade. The mid- to late-1970s saw electronic mail offerings as an improved version of telex done through public telephone lines. Then the pace quickened.

Electronic mail and videotex are two complementary solutions to the modernization of message services. Looking at these services from a fundamental stance we should appreciate that messages are communications, written or oral, sent between persons, and/or equipment.

They are formal or informal, involve text and data, and are handled by systems designed for processing, storage and transmission. Messages are informative, but after they have performed their function, they can be archived or dispensed.

Messages rarely alter the contents of a database in any other way than adding to it. By contrast, a transaction is a recoverable sequence of database actions. As such, it features three classes of elements: data processing, databasing, and datacomm.

Yet, while the fundamentals are practically agreed upon by everybody, opinions differ in terms of implementation. Electronic mail (the service) means different things to different people. To some, it is a buzzword used to indicate a computers- and communications-based interoffice mail system. To others, it is one of the pillars of office automation.

Electronic mail brings computers and communications resources to the office of every manager and every secretary—generally to every person able to use this facility. The rational approach is to look at electronic mail in broad terms to encompass:

- facsimile
- keyboard input
- voice-activated terminals
- other forms of information transfer

Electronic mail is a communications means that can link different computer-based devices together. At the same time, it is a communications discipline.

The protocols used in electronic mail help dissociate sender from receiver through store and forward capabilities. Therefore, they substitute in an able manner one-way telephone calls. It can handle digitized voice messages—while graphics and color capabilities are supported by videotex.

22.6 COMPUTER CONFERENCING AND DOCUMENT HANDLING

Computer conferencing goes beyond the now established Electronic Mail functionality. It *focuses on topics*—not individuals. An *electronic meeting* involves:

1. Video teleconferencing

2. Audio

3. Computer-based message exchanges

Computer conferencing integrates video capabilities with audio supports, offering: unlimited number of attendees, portable PC media, and the ability to locate individuals through computer search. Not all participants need to be online. This gives:

- *Time independence*

- *Distance independence*

- *Good journaling*

and uses supports already in place for Electronic Mail. Such connections can be: (1) one to one; (2) one to many; (3) many to many. With many to many, we allow groups to form and participate in electronic messaging.

Computer conferencing leads to the notion of a Virtual Conference Table (VCT). The computer/communications-based store and forward facility forms such a Virtual Conference Table. Around it, the participants hold their conference.

Each member can, as well, participate in other VCTs provided:

- Topics are well defined

- He can timeshare his personal attention

This is known as *branching*. It permits addressing different interest groups. *Joining* means getting on selected VCT and getting updated on subject treatment.

The environment must not grow indefinitely. This defines *boundary conditions*—which is the fourth critical factor in computer conferencing. Also from an administrative viewpoint the *inbox* of the participants must be carefully studied. Organization is important and involves:

1. Purpose

2. Duration

3. Participation

4. Access control

5. Meeting cycle(s)

6. Coordination effort

7. Adaptation to change

A further development in computer communications is the compound electronic document. The modelling effort of the *future office system* should focus on the *electronic document* as the fundamental object: This contains text, data (tables), charts, images, formulas, footnotes.

Documents can be: shown on video, printed on paper, sent over lines, and stored on a file server. Conversion operations must be provided to translate between the various formats.

When working on *compound electronic documents* we should keep in mind that information systems must be able to easily handle the large variety of objects:

1. Textual

2. Tabular

3. Algorithmic

4. Pictorial

Integration must see to it that dissimilar objects may be used together: Algorithms within pictures, pictures within text, text within a table. Ideally there should be: (1) A common underlying structural model, (2) Similar editing and specification languages, and (3) A uniform way to store, retrieve and display all objects.

These are goals assigned to a network architecture such as IBM's *DIA/DCA* and the CCITT X.400. Released in June 1983, IBM's *Document Interchange Architecture* (DIA) is designed to make feasible general information interchange across a broad spectrum of different information handling engines.

DIA is not a program product. It is a set of conventions that can be applied to a variety of products. The term *document* is used to refer to user-created information: data, text, graphics, image, voice, that flows in the network.

DIA defines protocols and data structures that subroutines use to communicate and exchange data. The other part of IBM's overall automation strategy is the *Document Content Architecture* (DCA).

1. The document content architecture *defines data formats that are compatible among dissimilar office systems*. Such compatibility allows all office systems whose design is based on this architecture to communicate with other offices in a transparent manner.

2. The document interchange architecture *defines functions for interchanging documents and other information between separate office systems connected through the network*. This permits users to: request document distribution and processing functions, address recipients, and retrieve documents from a library without having to know much about the physical organization of the network.

DIA and DCA describe the same notion of service but at a different level. Taken together, they allow a user to: enter a single request to distribute a document to multiple recipients, confirm delivery, and so on. As a parallel with the CCITT X.400 recommendation: DCA addresses itself to the task of *how to build a document*. Hence, it compares with the P2 protocol of the X.400 series. DIA tells *how to move a document*. This is done by the P1 protocol in X.400. In building a document, both DCA and the X.400 P2 leave open what the user can put inside a message. For instance, ASCII text.

As we can see there are alternatives in teleconferencing. The broad field of conferencing may be divided into seven classes:

1. Full motion videoconferencing
2. Freeze-frame videoconferencing
3. Computer conferencing
4. Shared screen computer conferencing
5. Electronic mail
6. Audio teleconferencing with graphics assist
7. Audio teleconferencing

As it has been already underlined, computer conferencing *focuses on topics*— not on individuals, offering: unlimited number of attendees, portable PC media, the ability to locate individuals through computer search. No doubt this range of facilities will be greatly expanding. *What is important is to look after a coherent introduction of such services—and this can only be done through knowhow, patience, and hard work.*

Appendix 1

Abbreviations in Modern Technology

ACK/NAK	Acknowledgment/No Acknowledgement
ADF	Advanced Development Facility
ADMD	Administration Management Domain (in X.400)
AI	Artifical Intelligence
ANSI	American National Standards Institute
AP	Application Program
ARB	IBM's Architectural Development Board (for SNA)
AS	Application System
ASCII	American Standard Code for Information Interchange
ATM	Automated Teller Machines
BAB	Baseband
BIOS	Basic Input Output System
BIU	Bus Interface Unit
BPS	Bits per Second
BPW	Bits per Word
BRB	Broadband
BSC	Binary Synchronous Communications
BSP	Business System Planning
CAD	Computer Aided Design
CAM	Computer Aided Manufacturing
CATV	Commodity Antenna Television. Cable Television
CBMS	Computer Based Message Systems
CCITT	International Telephone and Telegraph Consultative Committee

CEPT	Committee of European Post and Telecommunication (after which the 1983 Videotex standard was named)
CPU	Central Processing Unit
CRC	Cyclic Redundancy Check
CSDN	Circuit Switched Data Network
CSMA	Carrier Sensing Multiple Access
CSMA/CD	CSMA/Collision Detection
DB	Database
DBA	Database Administrator
DBMS	Database Management System
DC	Data Communications
DCE	Data Circuit-Terminating Equipment
DCF	Document Composition Facility
DES	Data Encryption Standard
DIA/DCA	Document Interchange Architecture/Document Content Architecture
DIF	Data Interchange Format
DISOS	Distributed Office Support System
DIU	Document Interchange Unit
DP	Data Processing
DRCS	Dynamically Redefinable Character Set
DSS	Decision Support Services (or Systems)
DTE	Data Terminal Equipment
EBCDIC	Extended Binary Code Decimal Interchange Code
ECMA	European Computer Manufacturers Association
EFT	Electronic Funds Transfer
EMail	Electronic Mail
EOT	End of Transmission
ESystem	Expert System
EUF	End User Functions
FDM	Frequency Division Multiplexing
FMP	Function Management Profile (of SNA)
FSM	Finite State Machine
4GL	Fourth Generation Language
5GL	Fifth Generation Language
GCD	Graphic Codepoint Definition
GKS	Graphics Kernel System
G-Sets	Graphic Sets

HDLC	High Level Data Link Control
HW	Hardware
ICST	Institute of Computer Science and Technology
IE	Information Element
IEEE	Institute of Electrical and Electronics Engineers
IGES	International Graphics Exchange Standard
IOCS	Input Output Control System
IPM	Interpersonal Messaging
ISoft	Integrated Software
ISO/OSI	International Standard Organization/Open System Interconnection
KB	Knowledgebank
KBPS	Kilobits per Second
KIPS	Kilo-Instructions per Second
LAN	Local Area Network
LAP	Line Access Protocol
LDDI	Local Distributed Data Interface
LU	Logical Unit (of SNA)
MAC	Media Access Control
MAC	Message Authentication Code
MAU	Media Access Unit
MBPS	Megabits per Second
MD	Management Domain (in X.400)
MHF	Message Handling Facility (in X.400)
MHS	Message Handling System (in X.400)
MIPS	Million Instructions per Second
MT	Message Transfer
MTA	Message Transfer Agents (in X.400)
MTAE	Message Transfer Agent Entity (in X.400)
MTL	Message Transfer Layer (in X.400)
MTS	Message Transfer Service
NA-PLPS	North American Presentation Level Protocol Syntax
NAU	Network Addressable Unit(s), (of SNA)
NBS	National Bureau of Standards
OA	Office Automation
OS	Operating System

PAC	Personal Authentication Code
PAD	Packet Assembly/Disassembly
PBX	Private Branch Exchange
PC	Personal Computer
PCB	Printed Circuit Board
PDI	Picture Definition Code
PEL	Logical Picture Element
PIN	Personal Identification Number
PIU	Path Information Unit (of SNA)
PIXEL	Physical Picture Element
PLP	Presentation Level Protocol
POS	Point of Sale Equipment
PRMD	Private Management Domain (in X.400)
PROFS	Professional Office System
PSDN	Packet Switched Data Network
PSL/PSA	Program Statement Language/Program Statement Analyzer
PSTN	Public Switched Telephone Network
PU	Physical Unit (of SNA)
QMF	Query Management Facility
RH	Request/Response Header (of SNA)
RU	Request/Response Unit (of SNA)
SBS	Small Business System
SDLC	Synchronous Data Link Control
SNA	System Network Architecture
SNADS	SNA Distribution Services
S/S	Start/Stop, asynchronous communications
SSCP	System Service Control Point (of SNA)
SSS	Solid State Software
SW	Software
TCAM	Telecommunications Access Method
TDM	Time Division Multiplexing
TH	Transmission Header (of SNA)
TSO	Timesharing Option
TSP	Transmission Subsystem Profile (of SNA)
UA	User Agents (in X.400)
UAE	User Agent Entity (in X.400)
UAL	User Agent Layers (in X.400)

VAN	Value Added Network
VCT	Virtual Conference Table
VDI	Virtual Description Level Interface
VLSI	Very Large Scale Integration
VTAM	Virtual Teleprocessing Access Method
VTX	Videotex
WP	Word Processing
WS	Workstation
XNS	Xerox Network System

Appendix 2

Definitions for Data Communications

Datacomm refers to end-to-end transmission of information other than voice, music, or video.

Data sources may be digital or analog.

Transmission handles any data source—with digital transmission a subset characterized by digital implementation.

A *digital signal* has a limited number of discrete states prior to transmission.

An *analog signal* varies in a continuous manner and can be seen as having an infinite number of states.

A *transceiver* is a transmitter/receiver.

A *modem* modulates the telephone transmission frequencies at the sending side of the message—and demodulates them at the receiving end.

The switching of frequency channels within a band is accomplished by *frequency agile* modems.

Baud is a unit of digital signaling rate equal to the reciprocal of the length, in seconds, of the shortest signal element.

The information rate in *bits per second* (BPS) may be greater than the baud rate, as one signaling element may represent more than one bit.

Erlang reflects the capacity of a communications system. It is a unit of traffic intensity, has no dimension, and is employed to express:

- The average number of connections under way
- The average number of devices used on a line

Traffic in erlang is the sum of the holding time of paths divided by the period of measurement.

One erlang expresses the continuous occupancy of one traffic path.

The *Bit Error Rate* (BER) is a measure of the performance of a digital transmission system. It expresses the probability of error per bit transmitted.

Word is a set of digital signs expressing information in an orderly, pre-established manner.

ASCII (AMERICAN STANDARD CODE FOR INFORMATION INTERCHANGE) is a 7-bit code, providing up to 128 different characters.

PLP also admits 8-bit ASCII (G-SETS, C-SETS).

Framing is the process of establishing a reference, so that the elements of a frame can be identified.

Frame is a segment of a signal (analog or digital) with repetitive characteristics, so that corresponding elements of successive frames represent the same thing. Each frame generally contains bits specified for source and destination addresses, control, data, and parity.

Transmission on a cable is *serial* in nature.

In *parallel* transmission the bits of a single message are sent at the same time, the whole message being transmitted in the time it takes to send one bit.

Nodes are attached to a network. The term identifies a basic information handling unit which is directly addressable.

Through *access techniques* nodes gain control of the common channel to transmit messages. Such techniques must be implemented in relation to a network topology.

Point-to-point connections by standard I/O channels limit distance and total connectivity while increasing cost and complexity.

A *multipoint* (MULTIDROP) link is a single line shared by more than two nodes.

A *bus* topology functions like a multidrop line, shared by a number of nodes. Bus architectures have most frequently been used for peer systems, with messages being broadcast to all nodes. Each node must be able to recognize its own address in order to receive a message.

Polling is a noncontention method of network access, determining the order in which nodes can take turns accessing the network.

Carrier Sense Multiple Access (CSMA) characterizes the ability of each node to detect traffic on a channel and act accordingly. This is a contention method.

Token is a bit pattern circulating around the carrier—whether physical or logical ring.

In a *ring* topology, the connected nodes form an unbroken circular configuration. A typical ring technology is a peer system of nodes—also serving as active repeaters for other nodes.

Rings with centralized control are referred to as *loops*.

Hub, or bridging point, is a device where a branch of a multipoint network is connected.

An *unconstrained* (hybrid, mesh) topology is non-specific taking various shapes and varying from one implementation to another.

The *network architecture* defines the relationship and interactions between attached devices.

Supported functions obey *protocols* and are served by common *interfaces*.

INDEX

Access problem, 77
Administration problem, 75
Advanced Research Projects Agency, 181
American Standards Association, 243
American Telephone & Telegraph, 85, 149, 187
Amplitude modulation, 272
Apple, 44, 45
Arpanet, 248
Artificial intelligence, 54, 83–85
Asset/liability management, 68
Automatic calling unit, 275
Automatic teller machine, 32

Bandwidth, 269
Batch processing, 36, 285–87
Bell Laboratories, 183
Benchmarks, 209–17, 219
Bennett, Dr., United Technologies, 57, 66
Bioelectronics, 153
Biotechnology, 153, 154, 163, 164, 167
Booz, Allen and Hamilton, study, 73
Bricklin, Daniel, 9
Buffer memory, 207

Cable TV, 34
CA-Executive, 25
Calendar, 75
Calendar services, 41
Carrier, 66
Cathode ray tube, 21
Charting, 11, 28
Citibank, London, 23–25
Coaxial cable connectors, 207, 267
Commodity software, 310
Communications facilities, 80
Communications services integration, 34
Communicating PC, 47
Compound electronic document, 328
CompuServe, 230
Computer aided design, 145, 146, 150, 189
Computer aided instruction, 207
Computer conferencing systems, 310, 327
Computer image processing, 28
Computer network, 312
Computer Technology Corporation, 147
Computer vision, 85
Concentrator, 291
Conditioning, 273, 274
Consultative Committee International for Telegraph and Telephone, 240, 274, 277
Control Data Corporation, 147
Cryptographic systems, 78
Cursor, 169

Databases, 78, 111
  organizational, 29
  public, 69
Database/datacomm system, 127
Database maintenance facilities, 130
Database management language, 25
Database management system, 100
Databasing, 33, 44, 104, 155
Datacomm, 33, 44, 104, 134, 155, 263–65, 280
Datacomm protocols, 49
Data dictionary, 39
Datamation, 49
Datapac, 321
Data processing, 32, 33, 104, 322
Decision support systems, 33, 87, 101, 102, 104, 107, 108, 111, 113, 141
Department of Defense, 167
Digital Equipment Corp., 84, 85, 207
Distributed database, 39
Distributed data processing, 42, 45
Distributed information systems, 45, 284
Document capture, 82
Document Content Architecture, 328, 329
Document handling, 112
Document Interchange Architecture, 328
Document management, 38
Dow Jones, 66
Dynamic routing, 315

Editing functions, 14
Eisenhower, Dwight, 161
Encryption capabilities, 17, 18
End user functions, 44, 89, 104
Electronic document, 328
Electronic files, 79
Electronic tellers, 124
Electronic typing aids, 75
EMail, 38, 47, 66, 75, 80, 310, 311, 326
Equalization, 272
Euronet, 321
Expert-Ease, 55
Expert systems, 84, 86–88, 92, 96–100

File handling, 30
Filing and retrieval systems, 75
Fixed routing, 312
Flexibility, 76
Flexible discs, 200–202
Foch, Marshall, 98
Fourth generation languages, 6, 302
Frequency division multiplexing, 290, 307
Frequency Modulation, 272
Frontend processor, 292
Full duplex, 278, 280

Gateways, 250, 324
General Electric, 84
Georgia Pacific, 128
Gilliland, Ezra T., 269
Graphic tablet, 36, 50
Graphics, 11
  computer, 14, 28
  design, 29
Grey, Mr. Chairman of the Board, United Technologies, 59

Half duplex, 278, 280
Hardware standards, 320
High level data link control, 250
Hitachi, 183, 207
Home banking, 124
Honeywell, 147

Icons, 12, 29
Info elements, 29, 38, 39
Infopage, 108
Information center, 307, 308
Information integrity, 315
Information providers, 324
Integrated circuits, 180
Integrated software, 173
Intel, 148
Intelligent terminals, 232
Interactive graphics, 54, 55
Interactive systems, 103
Interface program, 112
International Business Machines, 39, 49, 50, 85, 144, 149, 207, 301
  PC, 45, 55, 66
International Standards Organization, 249

Joystick, 50, 172, 173

Kilobits per inch, 195
Knowledge workers, 57, 62
Kodak, 194

Laser, 34
Laser-actuated memory devices, 34. (*See also* optical disc)
Laser optical technology, 207
Line controller, 276
Link, personal computer-to-mainframe, 17
Lisa, 172
Local area network, 27, 28, 33, 40, 46, 48, 49, 66, 88, 204, 283, 301, 302, 321
Logical interface, 231
Long haul communications, 17
Long range planning, 62
Lotus 1-2-3, 25

Macintosh, 9
Mailbox, 66
Mainframes, 305, 306
Maintenance, 52
Magnavox, 207

Management information systems, 26
Megabits per second, 35
Mellon Bank, 22
Memory management unit, 144
Menus, 104
Metaphors, 12
Microcomputers, 73
  communicating, 24
  distributed, 28
Microfile, 39, 79
Microprocessors, 35, 142–45, 156, 160
Microsoft, 115
Microwave, 268
MIT, 148
Mitsubishi, 40
Modems, 272
Modules, 90
Monitoring, 316
Morse, Samuel F.B., 268
Motorola, 45, 147, 148
Mouse, 15, 50, 171, 172
Multimedia networks, 31
Multiplan, 25, 45
Myriaprocessors, 40

NCR, 147, 207
National Computer Conference, Houston 1982, 21
National Semiconductor, 147
NEC, 189
North American Presentation Level Protocol Syntax, 245, 319
Nuttall, George, 161

Office automation, 7, 36, 54, 74
  definition of, 8
  system, 20, 73, 75, 79, 82
  system strategy, 23
Office system, 37, 38
Office system study, 81
Olivetti, 115
  M20, 115, 116
Open Systems Interconnection, 245
Operating system, 143, 219, 220
Oppenheimer, Robert J., 65
Optical discs, 34, 206. (*See also* laser-actuated memory devices)
Optical fibers, 34

Packet switching, 255, 256, 259, 300
Personal computers, 3, 9, 29, 44, 46, 52, 204, 237, 294, 295, 297, 298, 326
PC and mainframe connection, 297, 298, 300, 309
PC communications, 300
Phase Modulation, 272
Philips, 207
Phonemic Recognition, 190
Phonemic synthesis, 191
Physical interface, 231
Pigeonhole, 116, 120

Point of sales, 124
Price performance ratio, of computer hardware, 50
Private branch exchange, 149
Procedural standards, 320
Project teams, 25
Protocols, 248–51, 253
Prototyping, 6

Quality problem, 77
Query by form, 25

RCA, 207
Realtime, 286, 287
Removable discs, 199
Restart and recovery procedures, 316
Robotics, 146
Routing, 314
Routing pages, 110
Rubinstein, Arthur, 91

Sanyo, 189
Schlumberger, 84
Scrambling, 272
Simplex, 278, 280
Small business system, 27
Smalltalk, 55
Softcopy, 49
Software
    integrated, 10, 11, 28
Software standards, 320
Sony, 197
Speech recognition, 182, 190, 191
Speech synthesis, 189
Speech technology, 180
Speech understanding research, 181
Spreadsheets, 114–17, 122, 123
    calculations, 68
    generations of, 10
    matrix, 121
Star, 172
Storage Technologies, 207
Strategic Defense Initiative, 167
Streaming tapes, 204–206
Strowger, Almon B., 269
Supercomputers, 28
Supermicro computers, 9
Symphony, 9
Synchronous data link control, 250
Synthesized speech, 191
System integration, 8, 23
System project components, 37
System study, 220, 221–24

Telecommunications, 229
Teleprocessing, 283, 285
Texas Instruments, 183, 187
Text processing facilities, 80, 81
The Source, 230
Thomson CSF, 207
Time division multiplexing, 290
Time management, 112, 113

Timesharing, 289, 290
Toshiba, 189, 207
Touch sensing screen, 50, 175
Trackballs, 15
Transmission speed, 269
Transpac, 321
Transport problem, 77

United Services Automobile Association (USAA), 21, 22
United Technologies, 47, 58
Univac, 147
User interface, 74
Utterance Recognition Systems, 190

Value added networks, 133, 134, 267, 320, 321
Vertical recording, 202, 203
Video conferencing, 230
Videotex, 28, 32, 34, 44, 45, 47, 75, 82, 88, 286, 323–26
Virtual memory, 39
Visicalc, 9, 26, 59
Visicorp, 9
Visi-On, 25, 172
Visual programming, 29
Voice-actuated devices, 50
Voice annotation, 40
Voice entry, 35
    systems, 36
Voice input, 50
Voice input systems, 188
Voice mail, 193
Voice prints, 180
Voice recognition, 180, 182, 184, 187, 193
    speaker-dependent, 184
    speaker-independent, 184
Voice store and forward, 180, 181
Voice synthesis, 182
Von Neumann, John, 155

Waveform digitized speech, 191
Wide Area Telephone Service, 275
Winchester technology, 195–97
Windows, 15, 16, 173–75
Word processing, 3
Word processing commands, 60
Word processors, stand-alone, 21
Workstation, 46, 67, 130
    CAD, 141
    communicating, 14, 22
    concept, 12
    individual, 79
    intelligent, 24, 26, 29
    interactive, 41
    management, 66
    multifunctional, 5
    personal, 3–18, 41, 43
    physical, 4
    professional, 6, 22, 269

Xerox, 55, 207